Global Monitoring of
Terrestrial Ecosystems

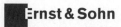

Global Monitoring of Terrestrial Ecosystems

Edited by W. Schröder, O. Fränzle, H. Keune and P. Mandry

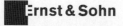

Editors addresses

Dr. Winfried Schröder
Geographisches Institut der Universität Kiel
Ludewig-Meyn-Straße 14
D-24098 Kiel

Die Deutsche Bibliothek – CIP-Einheitsaufnahme

Global Monitoring of terrestrial ecosystems / ed. by Winfried Schröder,
Otto Fränzle, Hartmut Keune and Patricia Mandry - Berlin: Ernst, 1996
ISBN 3-433-01533-3
NE: Schröder, Winfried [Hrsg.]

© 1996 Ernst & Sohn Verlag für Architektur und technische Wissenschaften GmbH, Berlin.

Ernst & Sohn is a member of the VCH Publishing Group.

All rights reserved (including those of translation into other languages). No part of this book may be reproduced in any form – by photoprint, microfilm, or any other means – nor transmitted or translated into a machine language without written permission from the publisher.

The quotation of trade descriptions, trade names or other symbols in this book shall not justify the assumption that they may be freely used by anyone. On the contrary, these may be registered trademarks or other registered symbols, even if they are not expressly marked as such.

Typesetting: SIB Satzzentrum in Berlin, Berlin
Printing: Alphabet KG, Berlin
Binding: Bruno Helm, Berlin
Production: Fred Willer, Berlin

Printed in Germany

Foreword

Government decisions related to the environment are based on far-reaching and responsible political concepts and focused on sustainable development in all areas of human activities. But at any level, local, regional or global, genuinely appropriate decisions can only be taken where there is sufficient and accurate knowledge of the environment. Scientists all over the world are trying to improve their understanding of the structures and functions of ecosystems, the fluxes of materials through them, and the interactions between different types of ecosystems. In addition, numerous monitoring, environmental information, and remote sensing programmes have been established to support decision-making at all levels. If it were possible to combine all the data and information from these different programmes in a meaningful way it might be possible to produce a reliable assessment of the world's environmental situation. Unfortunately, proper integration and aggregation of data is often not possible even on a small scale. We are confronted with a situation in which most of the data available cannot be used outside of the projects in which they were collected because they are not properly harmonized, i.e. they are neither comparable nor compatible.

Based on an initiative of the G-7 countries in 1987, the German Government offered to host and support a UNEP office under the auspices of the Global Environment Monitoring System (GEMS) to look into ways of improving harmonization of environmental measurement. The UNEP Harmonization of Environmental Measurement Office (UNEP-HEM) was established in 1989 in Munich on the campus of the GSF-Research Centre for Environment and Health. The HEM mission is to enhance the compatibility and quality of information (enhancing harmonization) on the state of the environment world-wide.

In order to work consistently towards harmonization of environmental observations it is necessary to have a conceptual framework describing how, where and by whom data is collected, integrated, aggregated and assessed, and indicating the cross-links and relationships between different, often apparently disparate, activities.

The United Nations, in particular UNEP, UNESCO, WMO and FAO, and the International Council of Scientific Unions (ICSU) have established Global Observing Systems for Climate (GCOS) and Oceans (GOOS) and are in the process of establishing a Global Terrestrial Observing System (GTOS). These global observing systems will help to improve the current situation by providing the organisational framework needed to promote and ensure the collection of comparable and compatible data and information on the environment world-wide. They will also provide a potentially valuable basis for developing the conceptual framework needed for developing a coherent approach to harmonization activities on a wider scale.

The Federal Ministry for Education, Science, Research and Technology (BMBF) encourages any activities aimed at increasing the potential usefulness of the data acquired within different projects and reducing redundancy and overlap as it is hoped that such activities will help to maximise effectiveness. For this reason we supported HEM's suggestion of calling together a group of international experts to initiate, discuss and stimulate the development of a conceptual framework for environmental observation as a basis for developing a coordinated approach to harmonization. Specifically, the BMBF provided the support to hold a workshop on 'Harmonization of Environmental Data' which was closely linked to the GTOS planning process.

It is hoped that our support of the workshop has helped to provide the forum needed to make a step forward towards improved harmonization of environmental observation and thus to a better understanding of the state of the environment world-wide.

Dr. Peter Krause
Federal Ministry for Education, Science, Research and Technology
Head of Division for Global Change

Contents

Foreword V

Annexes IX

List of Contributors XI

1	**Introduction**	1
1.1	Rationale, aim and objectives of the book	2
1.2	Acknowledgements	
	W. Schröder, O. Fränzle, H. Keune and P. Mandry	3
2	**Towards a Global Terrestrial Observing System (GTOS). The UNEP-HEM workshop on Harmonization of Environmental Data, 7–10 June, 1994. Presentations and written contributions**	3
2.1	Background	3
2.1.1	Environmental information - autobahn or maze?	
	E. F. Roots	3
2.1.2	Earthwatch: An introduction to the UN system-wide coordination of environmental observation	
	A. L. Dahl	33
2.1.3	GTOS – another step toward understanding the earth system	
	M. F. Baumgardner	37
2.1.4	GTOS National needs information requirements	
	H. Narjisse	43
2.1.5	Harmonization of environmental data: the requirements for developing a consistent view of the environment world-wide	
	H. Keune and P. Mandry	51
2.2	Disparities in sampling, parameters and metadata: environmental monitoring and assessment as a unifying basis	
	W. Schröder and O. Fränzle	57
2.3	Example for, and contribution to GTOS at global, regional and national levels	67
2.3.1	Data management in ocean sciences	
	D. Kohnke	67
2.3.2	Quality assurance/quality control as prerequisite for harmonization of data collection and interpretation within the WMO-Global Atmosphere Watch (GAW) Programme	
	V. A. Mohnen	77
2.3.3	Terrestrial monitoring in the Arctic environment	
	L.-O. Reiersen	85
2.3.4	Integrated environmental monitoring network development: the experience of the Environmental Change Network (ECN) (United Kingdom)	
	A. M. Lane	91

2.3.5	The Swiss concept of integrated ecosystem-related environmental observation: the ecosonde *K. Peter and P. Knoepfel*	97
2.3.6	Some principles for the organisation of GTOS *J. Nauber*	103
2.3.7	Challenges for developing countries *L. A. Ogallo*	105
2.3.8	The Russian view on forming GTOS *Y. A. Pykh*	107
2.3.9	Possible contribution of the Nature Reserves Authority (Israel) *J. Cohen*	111
2.4	Constraints to the extraction of information from complex environmental datasets *I. K. Crain*	113
2.5	Model-supported synthesis, evaluation and application of environmental information *H. F. Kerner*	123
2.6	Integrated assessment concept of the working group „Informationsystems for environmental research and planning" at the GSF *R. J. M. Lenz, M. Knorrenschild, M. Schmitt and R. Stary*	133
2.7	Estimating vertical fluxes of gases between atmosphere and ecosystems *U. Dämmgen, L. Grünhage and H.-J. Jäger*	145
2.8	Environmental-economic accounting: how can it support decision-making *W. Radermacher*	155
2.9	Standard reference data and policy assistance systems for global health evolution *C. Greiner, F. J. Radermacher and T. M. Fliedner*	159
2.10	Workshop results	165
2.10.1	Impressions at the close of the workshop *E. F. Roots*	165
2.10.2	Summary report of the workshop *M. F. Baumgardner*	183
3	**Proposal for a global concept for monitoring terrestrial ecosystems as a basis for harmonization of environmental monitoring** *O. Fränzle, W. Haber and W. Schröder*	195
4	**Summary** *H. Keune, P. Mandry, W. Schröder and O. Fränzle*	207

Annexes 213

ANNEX I 213
Participants in the Workshop on Harmonization of Environmental Data,
7–10 June 1994, Munich, Federal Republic of Germany 213

ANNEX II 219
1   Global Climate Observing System (GCOS) 219
2   Global Ocean Observing System (GOOS) 224
3   Terrestrial Observing Systems of Global or Regional Relevance 226
3.1   EuroMAB Biosphere Reserve Integrated Monitoring
        (BRIM) for Biological Diversity Conservation and
        Sustainable Use of Natural Resources 226
3.2   Integrated Monitoring by UN-ECE 239
3.3   European Environment Agency (EEA) 245
3.4   Arctic Monitoring and Assessment Programme (AMAP) 250
3.5   Environmental Monitoring and Assessment Programme (EMAP), USA 255
3.6   Long-Term Ecosystem Research (LTER), USA 262
3.7   National Oceanic and Atmospheric Administration:
        Mussel Watch Project, USA 267
3.8   Terrestrial Ecosystem Research Network (TERN), Germany 269
3.9   Chinese Ecological Research Network (CERN), China 274
3.10   Environmental Change Network (ECN), UK 278

List of Contributors*

Marion F. Baumgardner
Purdue University, Department of Agronomy, West Lafayette, IN 47907-1150, USA

Josef Cohen
Nature Reserves Authority, Head Land & Information Unit, Division of Science and Management, 78 Yirmeyahu St., Jerusalem 94467, Israel

Ian K. Crain
The Orbis Institute, P.O. Box 20185, Ottawa, Ontario K1N 9P4, Canada

Arthur L. Dahl
Coordinator, UN System-Wide Earthwatch, EARTHWATCH Secretariat, UNEP, P.O.Box 356 CH 1219 Chatelaine Geneva, Switzerland

Ulrich Dämmgen
Institut für agrarrelevante Klimaforschung der FAL, Eberswalderstr. 84F, D–15374 Müncheberg, Germany

Otto Fränzle
Projektzentrum „Ökosystemforschung Bornhöveder Seenkette", Geographisches Institut der Christian-Albrechts-Universität, Schauenburgerstraße 112, D–24118 Kiel, Germany

C. Greiner
Central Institute for Biomedical Engineering, University of Ulm, Albert Einstein Allee 11, 89070 Ulm, and Institute for Clinical Physiology, Occupational and Social Medicine, University of Ulm, Albert Einstein Allee 11, D–89081 Ulm, Germany

Ludger Grünhage
Institute for Plant Ecology, University of Gießen, Heinrich-Buff-Ring 38, D–35392 Gießen, Germany

Wolfgang Haber
Technische Universität München, Lehrstuhl für Landschaftsökologie, Hohenbacherstr. 19, D–85350 Freising-Weihenstephan, Germany

H. Franz Kerner
TEAM 18 – Umweltforschung, Bahnhofstr. 18, D–85354 Freising, Germany

Hartmut Keune
UNEP-HEM, c/o GSF Neuherberg, P.O. Box 1129, D–85758 Oberschleißheim, Germany

---

* For contributions by three authors or more, only the first author is listed.

Peter Knoepfel
Institut de hautes études en administration publique (IDHEAP). Route de la Maldìere 21, CH–1022 Chavannes-près-Renens, Switzerland

Michael Knorrenschild
GSF Projekt Umweltgefährdungspotentiale von Chemikalien, Neuherberg, Postfach 1129, D–85758 Oberschleißheim, Germany

Dieter P. Kohnke
Bundesamt für Seeschiffahrt und Hydrographie (BSH), Bernhard-Nocht-Str. 78, D–20359 Hamburg, Germany

A. Mandy Lane
Institute of Terrrestrial Ecology, Merlewood Research Station, Windermere Road, Grange-over-Sandes, Cumbria LA11 6JU, U.K.

Roman J. M. Lenz
GSF Projekt Umweltgefährdungspotentiale von Chemikalien, Neuherberg, Postfach 1129, D–85758 Oberschleißheim, Germany

Patricia Mandry
UNEP-HEM, c/o GSF Neuherberg, P.O. Box 1129, D–85758 Oberschleißheim, Germany

Volker A. Mohnen
State University of New York, University at Albany, c/o ASRC, 100 Fuller Road, Albany NY 12205, U.S.A.

Hamid Narjisse
Institute Agronomique et Veterinaire, Hassan II, BP. 6202, Rabat, Morocco

Jürgen Nauber
German Man and the Biosphere (MAB) National Committee, c/o BfN, Konstantinstr. 110, D–53179 Bonn, Germany

Henry Nix
The Australian National University, Centre for Resource and Environmental Studies, Canberra ACT 0200, Australia

Laban A. Ogallo
National Council for Science and Technology, Emperor Plaza, 2nd floor, P.O.Box 30623, Nairobi, Kenya

Kathrin Peter
Schweizerische Kommission für Umweltbeobachtung, Projektleitung und Koordinationsstelle, Bärenplatz 2, CH–3011 Bern, Switzerland

Yuri A. Pykh
Center for International Environmental Cooperation (INENCO), Russian Academy of Sciences, 4 Chernomorsky per., St. Petersburg 190000, Russia

F. J. Radermacher
Forschungsinstitut für anwendungsorientierte Wissensverarbeitung (FAW), Universität Ulm, P.O.Box 2060, D–89010 Ulm, Germany

Walter Radermacher
Statistisches Bundesamt, Abteilung IV, D–65180 Wiesbaden, Germany

Lars-Otto Reiersen
General Secretary Arctic Monitoring Assessment Programme, P.O.Box 8100 Dep., N-0032 Oslo, Norway

E. Fred Roots
Department of the Environment, K1A OH3 Ottawa, Ontario, Canada

Winfried Schröder
Projektzentrum „Ökosystemforschung Bornhöveder Seenkette", Geographisches Institut der Christian-Albrechts-Universität, Schauenburgerstraße 112, D–24118 Kiel, Germany

1 Introduction

W. Schröder, O. Fränzle, H. Keune and P. Mandry

1.1 Rationale, aim and objectives of the book

Many environmental problems are now recognised to be essentially global in nature, i.e. either occurring on a global scale, or sufficiently widespread to have planetary influence or significance. Increasingly decisions are having to be made which require information from a large number of disparate sources, across a multitude of disciplines and covering a broad geographic scale. Data from different sources, however, can only be integrated and aggregated meaningfully if they are comparable and compatible and of known and defined quality, i.e. the data must be harmonized. Harmonization must ensure not only that data collected in different programmes are comparable and compatible, it should also ensure that integration and aggregation of data are performed at different places in such a way that the results are still comparable at the level of environmental information, and that assessment is done in a comparable fashion so that a common understanding about the state of the environment can be developed at all levels. Harmonization is thus a tool for ensuring that data and information from different sources can be meaningfully combined.

The UNEP-Harmonization of Environmental Measurement (HEM) office, as part of UNEP's Environmental Assessment Sub-Programme, has been considering problems of harmonization in environmental monitoring and environmental information management for more than four years. HEM has identified two prerequisites for genuinely useful cross-sectoral harmonization activities: a conceptual framework for the monitoring activities, and a comprehensive and consistent catalogue of information on the players/contributors to the overall endeavour.

One way of developing a consistent picture of the environment on a global scale, i.e. obtaining (sufficient) comparable and compatible data across the globe, is to establish effective global networks to monitor and detect current status, trends and change in environmental systems using representative sites or areas, or data collected from different sources in a harmonized fashion. Recently, UNEP, UNESCO, WMO, FAO and ICSU have established an Ad Hoc Scientific and Technical Planning Group to develop a basic concept for a Global Terrestrial Observing System (GTOS). The GTOS activities will cover all aspects of terrestrial environmental monitoring and assessment. HEM considers that GTOS could, or should, provide a focus for global cross-sectoral harmonization within the area of environmental monitoring, as well as providing a real practical basis for developing a framework for further harmonization efforts. Equally, one of the keys to the success of GTOS will be the development of effective integrated measures for harmonization of data collection, integration and assessment activities for the network itself. All of this is in the centre of the activities of the HEM office.

For GTOS to be successful, it will be necessary to develop strategies for the timely and efficient flow of credible, useful environmental data along the path from the time of data acquisition to information delivery. In order to know what relevant data, or sources of data, already exist it will be important to know who is observing or monitoring what, where, when, how and why. Cross-sectoral or integrated monitoring will be required to obtain a complete picture of the state of an ecosystem, and a plan will be needed showing the requirements for such an integrated approach, the links between different data sets, and the means for ensuring the comparability and compatibility of data. Thus GTOS offers a means of developing and testing, on a limited scale, both a conceptual framework for terrestrial ecosystem assessment, and a comprehensive catalogue of information.

HEM has supported GTOS planning activities from the beginning, and is particularly interested in assisting the necessary harmonization activities within GTOS. With this in view, HEM organised a workshop on harmonization of environmental data with specific reference to GTOS planning activities.

The UNEP-HEM workshop on Harmonization of Environmental Data took place on the campus of its host institute, the GSF-Research Centre for Environment and Health, in Munich on 7–10 June 1994. The meeting was sponsored by the German Federal Ministry of Education, Sicence, Research and Technology (BMBF). It was attended by members of the Ad Hoc Scientific and Technical Planning Group for GTOS, (in particular members of Working Group I on Data Management, Access and Harmonization), representatives of UNEP-EAS and the GTOS Planning Group Secretariat, and individuals from different institutions (Annex I). The objective of the workshop was to discuss harmonization of environmental information in general and the requirements of GTOS in particular, outlining the different approaches and constraints involved in harmonization of environmental data collection, integration and analysis.

This book is based essentially on the proceedings of the workshop. It is intended to provide the reader with an overview of the basic problems involved in the harmonization of environmental data, the need for and potential benefits of a system such as GTOS, and the practical implications of harmonization for a global network. The different approaches used at present are indicated in a series of descriptions of a representative sample of ongoing programmes.

The book is divided into four chapters. Chapter 1 is introductory and gives an overview of the need for harmonization in general. Chapter 2 has three parts: Part 1 describes the current state of development of GTOS, Part 2 and 3 refer to the workshop on harmonization of environmental data. They contain a series of contributions describing different constraints to harmonization and discussing national, regional and global approaches to harmonization of environmental information. Part 3 closes with the summary report of the workshop results. Chapter 3 is a proposal for a global concept for environmental monitoring as a basis for harmonization. Finally, the book closes with a summary. A selection of ongoing programmes are described in Annex 2, providing a background for understanding the complexity of the problem.

References

HEAL, O. W.; MENAUT, J. C.; STEFFEN, W. L. (eds), (1993): Towards a Global Terrestrial Observing System (GTOS): detecting and monitoring change in terrestrial ecosystems. – Paris, Stockholm (MAB Digest 14 and IGBP Global Change Report 26)
UNEP (1994): Report of the First Meeting of the Ad Hoc Scientific and Technical Planning Group for a Global Terrestrial Observing System (GTOS) 13–17 December, 1993, Geneva Switzerland. – Nairobi (UNEP/GEMS Report Series No. 24)

1.2 Acknowledgements

Funding this project was provided by the German Federal Ministry of Education, Science, Research and Technology (BMBF) and is greatfuly acknowledged. Both the workshop and the preparation of this book benefited immensurably from the hard work and enthusiasm of all participants of the workshop on Harmonization of Environmental Data and all members of the HEM office.

2 Towards a Global Terrestrial Observing System (GTOS): The UNEP-HEM workshop on Harmonization of Environmental Data, 7–10 June, 1994. Presentations and written contributions

2.1 Background

2.1.1 Environmental information – autobahn or maze?
E. F. Roots

The Role of Information in the Evolution of Societies

The human species, characteristically self-centred and arrogant, has been fond of attributing its biological success and its present dominant position among animals on this planet to its ability, as a species, to possess and use information. We have even called ourselves *Homo Sapiens* – the "knowing man". Whether or not our present abundance in numbers and our ability to influence other species – mostly by destroying them – and to modify the planet on which we live is due to our mental abilities is a matter of some debate among anthropologists and biologists; but there is no doubt that *Homo Sapiens* differs from other species, as far as we can tell, by a greater facility in communicating memory and experience between individuals and in groups, in an ability to form and describe abstract ideas, and in the development of means to retain or store thoughts, experiences, and awareness of cause-effect relationships outside or beyond the memories of single individuals. In short, the human race operates largely on *information*, which is based mainly on individual observations that are collectively and systematically organized into *communicable knowledge*.

Throughout history, the way that the human race has accumulated and organized its information to develop knowledge and used it as a basis for decisions or actions has had much to do with the structure and nature of societies. Ernest Gellner[1] has pointed out that any group of humans that organizes itself to be more effective as a group than the individuals could be if they acted independently must develop relationships between three characteristics of any group or society – cognition, coercion, and production.

Cognition – the knowledge and awareness possessed by each individual or available to the group collectively;

Coercion – the ability for the decisions of a few to affect many, so that actions can be taken in concert through an established organization or structure (leadership, authority, management), and the mechanisms by which each individual can influence the central decisions or the collective action;

Production – the result of individual or collective actions on the natural world to generate or provide materials or means to satisfy physical or intellectual wants.

The achievement of each of these three elements requires a continually increasing supply of information. But different societies use their information in quite different ways. As a background to our discussions at this workshop on the problems of providing usable or compatible information about any or all aspects of the environment to the "user" or "decision-maker" for future decisions that are intended to improve human well-being and minimize or halt the progressive destruction of planetary life support capacity, it is useful to reflect for a moment on how environmental information has been used to support the societal characteristics of cognition, coercion, and production. If we in the science community are claiming that we are adapting our information and data systems to meet the needs of the user or decision-maker, we need to be as clear as possible who the users are, and what decisions will require our information.

Very briefly, and as a gross over-simplification, it can be noted that in *hunter-gatherer societies,* in which humankind has lived through nearly all its time on Earth and which may still be found on every inhabited continent, emphasis is on development of wide and acute powers of observation and knowledge about the environment at the individual level, with production, based on such knowledge and individual skills, tied to immediate needs. Personal cognition, based on direct observation and unified by abstract concepts passed on through generations and shared among immediate family units, is dominant; coercion is minimal, a tool for immediate production.

When, about 5000 years ago, humans began to organize to undertake advance actions to manipulate natural processes to provide sustained food, tools, weapons, and wealth, the role and situation of environmental information in society changed dramatically. The agrarian societies developed a structure, a division of labour and activity that resulted in centralized power, in the acquisition of wealth and means to control or protect it, the ability to expand social units to cities, nations, and empires, and a stratification of the means of coercion (state, church, military), production (farmers, trades people) and cognition (educators, clerisy). In such societies coercion, as a means to achieve stability and control production and behaviour, necessarily becomes dominant, and cognition is a sub-set of coercion and production. Information about the natural world and its resources is important to the degree that it is useful to maintain the socio-political system and economic production.

Beginning about the eighteenth century in western Europe, an offshoot of the agrarian society appeared which has made use of a continually developing technology whose purpose is to use the processes, materials and energies of Nature for accumulation of wealth and influence: the so-called industrial society. This societal structure has spread throughout much of the world. In such a system, production is dominant; social stratification and coercion is based on information and economic instruments, and not mainly on political structure, and social hierarchy is determined largely by continual competition. Cognition becomes increasingly technological, rather than conceptual, collective rather than individual. Information about the natural world serves a mainly economic purpose.

In the past three or four decades there has been emerging in many countries a new and still different structure of society, which is information driven, and in which production and coercion are tied to continual innovation, with rapidly changing life styles, mobility of occupations, where social cohesion is lessened, and established concepts disregarded in favour of a rapidly changing knowledge base. This has been called the post-industrial or information society. This societal philosophy, not held by the majority but widespread and growing throughout the world, has many similarities to the pre-agrarian hunter-gatherer societies; the technologies used and type of information are of course entirely different but the inter-personal ethics and dominance of individual cognition and action, with a disregard of established coercion, have many features in common.

It should be noted that these different societal structures have nothing to do with the form of government – democratic, socialistic, totalitarian, etc. – that is the focus of much political classification and concern. Such governmental forms are variations in the way that coercion and production are achieved within the agrarian and industrial societal structures, and have little to do with their dominance or relation to cognition. Thus the most rigid totalitarian state can make use of modern industry or natural knowledge to achieve its purpose, just as can the most liberal-minded free democracy.

Today, in most of the developing and developed world, there is an agrarian-style government structure, with decision processes based on an ethic that it is the role of government to maintain social order and the means of production of wealth through ever-expanding use of natural resources, or through use of resources more effectively, with central responsibilities for the control or removal of obstacles that threaten or constrain that expansion. The obstacles can include environmental deterioration. For that reason, the inhabitants of most countries assume it to be "natural" that the government should be responsible for preserving the quality of the environment and management of resources, even though government actions or policies may not have been responsible for the problems. The decision-making processes in an agrarian-style government are structured to maintain this system through coercion (laws, taxes, regulations, subsidies, administrative departments, more drastic action if necessary), with only such investment in cognition (education, research, monitoring) as may be necessary to meet those ends. However, the economic and production system in the developed world is dominantly in the industrial society mode, where success depends on competitive innovation and testing in the market-place, and not on maintenance of the established agrarian-type control system. The ethic of the industrial society is often in conflict with that of the power and control system within which it must operate. Industrial decision-making has typically quite different goals, focused on production for profit and individual well-being, with coercion and cognition subordinate and justified as a means to obtain "practical" results. The demand for, and use of, environmental information likewise tends to be focused on production, not control.

These established societal systems have made possible an increase of collective material wealth, and resulted in an ever-expanding amount of objective and subjective information, with development of methods to obtain it, and communicate it. They have led also to an increasing technical sophistication of millions of individuals, and thus have made possible the emergence of the so-called post-industrial or information society. The information society appears to be characterized by quite a different ethic and decision-making process from those of the agrarian and industrial societies from which it has come. Interchangeable knowledge without limit, disregard for "the established rules of the game", and power based on technical information and know-how only loosely connected to wealth or societal structure (a smart kid with a computer can jam the stock market) are characteristics of this post-industrial phase. The goals for decision-making are likewise different, much less clear, and tend to reject or disprove established concepts. Information itself becomes a source of influence, rather than a means to political strength or wealth.[2] And the attitude to nature and environment is changing dramatically, from one which the natural world has been seen as a storehouse of resources to be exploited for human gains[3] to a view that regards environment as the setting which supports humans and all other creatures and determines their well-being. Environmental information is valued as knowledge in its own right and as proof of the need for action to protect Nature or regulate human activities.[4]

This situation, plus the expansion of population beyond the requirements of the established system to "need" each human for coercive control or for production of food and wealth, has encouraged or driven a growing portion of younger members of society in both non-industrialized and technologically advanced parts of the world into an information-based hunter-gatherer mode. What else, for example, is Internet?

The established agrarian political structures and industrial systems, and also parts of the public at large, have become impressed, excited, and not a little bemused by the rapid emergence of information techniques and the almost limitless accessibility to technical information and intercommunications now possible among the technologically sophisticated. "If only this could be managed or controlled for our purposes", they say in effect, "we will have a tool of remarkable utility, one that should be able to counter the breakdown of society and economy as human population expands, natural resources become scarcer and we approach the limits of the ecological tolerance of the planet".

From this kind of thinking has emerged popular terms such as "information highway", or "information autobahn to the future". The hope for the future, it is said by some with great enthusiasm, is to get on the information highway; if one doesn't, one will be left behind. But this seems to said without much thinking about where the highway may be leading us, or who has gone ahead to survey the route and identify the destination. I do not know of any autobahn in Germany or any other country that was built by the travellers as they joined it and drove at high speed along it without a knowledge of where it was taking them. With the enormous increase of items and kinds of information, and the proliferation of technologies to assemble, interpret or disassemble such information, the "information highway" could be one of endlessly confusing interchanges, traffic gridlock; or it could lead many of us into a maze, from which the way out could be very obscure indeed.

That is why I have titled these introductory comments to our workshop with a question and an implied warning. Are increasing floods of environmental information and information-rich sophisticated monitoring systems contributions to the information autobahn to the future, or are they in danger of leading us into a maze, characterized by our being overloaded with information whose significance we do not know, inhibiting our positive or carefully considered actions and from which escape will be difficult?

Much will depend on how we "harmonize" the information to make it coherent and useful to someone, and how well those of us responsible for producing the information understand and meet the respective needs of the quite different kinds of decision-makers, in different components of our society, who are going to use our product for their own quite different purposes.

Data. Information, Knowledge

In this workshop we do not want to get delayed or diverted by definitions and arguments about terms. But I think it is necessary, especially in an international meeting where many are using a language that is not their mother tongue, for us to have a common concept of the ideas we have in mind when we talk about "measurement", "data", "information" and "knowledge".

I think most of us can agree that what we know, or think we know about the environment and natural phenomena comes from observations that we have made or have been made by others and communicated to us; and that those observations combined with many others give us information which is the basis of our personal or collective knowledge. If the observations are quantitative, recorded against some agreed standard that is independent of how we feel at the time and independent of what will happen to the result, we call those observations measurements, and we say that measurement is, or should be, as objective as we can make it. But when we begin to combine those measurements and give them more meaning, we have to add a purpose, a selected methodology and some judgement of quality; and the result becomes increasingly and properly, partly subjective.

A good way to visualize what are will be talking about in this workshop is to use the philosopher's "Staircase of knowing" (fig. 2.1.1.1)

On this staircase,
Observations and measurements, when verified according to agreed standards become data;
Data, properly selected, tested and related to subject areas can become *information;*
Information, organized and interpreted or applied to areas of interest or concern, can become *knowledge;*
Knowledge, if assimilated and subjected to mental assessment and enrichment, so that it is comprehended and integrated into a base of facts and impressions already assimilated, leads to *understanding.*
And understanding, put into perspective with judgement according to some values, can become *wisdom.*
Note that as one moves up the staircase, the material and ideas become increasingly subjective, with increasing human value added.[5]

At this workshop we are, I believe, concerned with the first three steps of this staircase. These steps are the basis and starting point for society's knowledge and understanding of the environ-

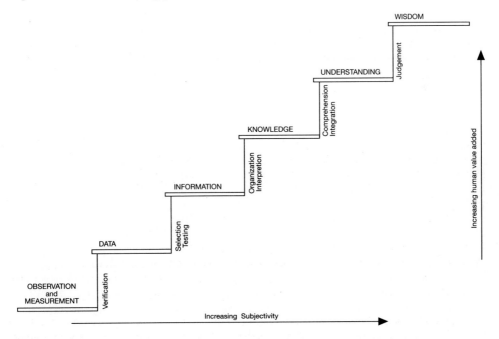

Fig. 2.1.1.1:
The Staircase of "Knowing"
(from Roots, E. F. [1992]. "Environmental Information – A Step to Knowlegde and Understanding". Environmental Monitoring and Assessment, No. 50, 4. Kluwer Academic Publishers, Dordrecht, Netherlands, pp. 87–94.)

ment and what is happening to it. But here we have a problem. Our stated purpose for obtaining and improving the measurements and data, to which the harmonization here discussed is an essential contribution, is to make the information available to the public and to authorized decision-makers so that better decisions can be made. But on what step of the staircase do the decision-makers stand? The increasing emphasis placed on more and better information can all too easily lead to an assumption or perhaps even a belief that if only a person had enough data or information, that person would somehow make wise decisions; and that technical know-how and ability to recognize or simulate consequences can replace value and judgement. Those of us immersed in the problems of environmental measurement, data, and information handling must not ignore the vertical "risers" on the staircase. Without stepping up, one can never step ahead to knowledge or understanding.

In the past, society and the environment have suffered greatly from policies and decisions that were based on dogma or on hierarchy or greed without adequate information about the consequences or the alternatives. In the 1990's we are perhaps in danger of equal or greater damage from policies and decisions that are driven by data and information without understanding, rather than driven by value and judgement based on good information.

Where does science fit on this staircase? Clearly, on every step. I do not want to carry the staircase analogy too far; but we might think of science and research as a sort of railing that helps progression from step to step, and sometimes provides a short-cut to get down to the bottom when one is stuck halfway up. For the staircase must be descended as well as ascended. The "measurements" with which we are concerned in this workshop will only be useful if:

- One measures the right things;
- One measures them properly according to tested and accepted standards and procedures;
- One measures them reliably and consistently;
- One conveys the results effectively.

Note that all of these qualifications for good measurements are subjective, arising from considerations higher on the staircase and conveyed through feedback loops from science, or from users or decision-makers on other steps. And it is in this feedback process that the need for harmonization is most evident.

What is Environmental Measurement?

In its simplest form, an environmental measurement is a quantitative representation of a characteristic or process of the natural world, in relation to some previously defined or accepted description used as a basis for comparison.

To achieve measurement of characteristics and processes of the natural world, it is necessary that:
- The feature or process of attention be identified and recognized separately from its surroundings wherever it may occur (but preferably recognized within its context);
- There is deliberate action to ascertain its amount or change;
- There is a widely recognized and preferably tested standard or basis of comparison.

These simple requirements are not always easy to satisfy. But without them, there cannot be reliable environmental measurement, monitoring, or data.

One problem, of course, is that environmental phenomena extend over the full range of time and space of natural characteristics and processes (fig. 2.1.1.2).

If one superimposes on this natural scheme the effects of human population and human activities (mostly on the "minute" to "century" time scale, but operating on all space scales up to global), it is clear that no one single measurement scheme can be applied. But if the measurements are to be harmonized, they must have a consistent philosophy. It was to identify and establish practices that would help achieve such a philosophy that in part, HEM was established; and this is one reason why we are here discussing the issues of harmonized measurements in the context of global observing systems.

International Recognition of Issues of Environmental Measurements

As background to our discussions on information requirements, data collection and data management, it may be useful to recall some of the developments that have led to the present situation with respect to environmental information, and its institutionalization.

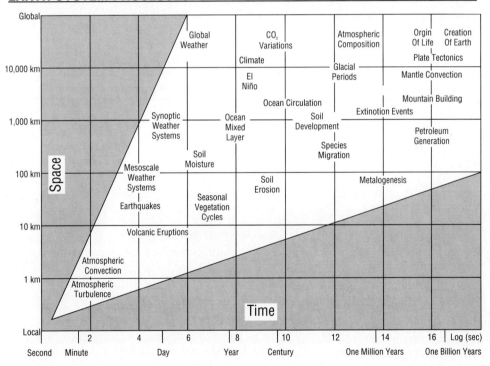

Fig. 2.1.1.2:
Earth System Processes: Characteristic Space and Time Scales
(from Bretherton, F. [chair] [1986]. "Earth System Science, a Program for Global Change". *Earth System Science Committee, NASA Advisory Council,* National Aeronautics and Space Administration, Washington, D.C., U.S.A., p. 48)

Although scientists from the beginning of organized scientific knowledge – Aristotle, Bacon, Galileo and Torticelli – had accurately described and recorded individual natural phenomena, it was not until the establishment of a network for systematic weather observations in 1819 that the importance of organized regular measurements of separate but related characteristics of "Nature" became widely recognized. The weather network was followed in 1844 by establishment of the association for simultaneous measurement of magnetic phenomena the (Magnetische Verein)[6] throughout most of Europe, which developed very high standards for precise observations and data. By the middle of the 19th century, there was a great blossoming of enthusiastic observation and description of every characteristic of Natural History, and many official instructions were written about how to make the observations and measurements compatible and interchangeable. The British "Admiralty Manual of Scientific Enquiry: Prepared for the Use of Officers in Her Majesty's Navy, and Travellers in General" published in 1849 and revised in 1851, is an outstanding example.[7] It is a remarkable and comprehensive scientific treatise devoted to compatible and harmonized observation and measurement in the astronomical and geophysical sciences, meteorology, hydrology, oceanography, geography, geology, zoology, botany, ethnology, and medicine; and it contains the strict statement:

> "A bad observation, or an observation where results are given without means of verification, is worse than no observation at all".

In many ways, our discussions to ensure harmonization throughout the Global Observing Systems are an attempt to up-date, after 150 years, the world-wide goals of the "Admiralty Manual of Scientific Enquiry".

A major step toward harmonization of environmental observation and data was of course made by the International Polar Year (IPY) 1882–83, and especially by its promoter and organiser, Karl Weyprecht, who campaigned tirelessly throughout Europe for rigorous attention to accurate and reliable observation and recording.[8] We need not concern ourselves here with the many political, scientific, and adventurous aspects of the IPY, or with profound effect that it had on internationalization of science and establishment of a peer review system to ensure scientific quality; but we should note the emphasis that this world-wide endeavour placed on careful and compatible environmental measurements (fig. 2.1.1.3).

These four statements are from a list of "principles of scientific observation" that Weyprecht drew up in 1879.

That 14 simultaneous expeditions to the polar regions and 39 established observatories in 25 countries throughout the world could take synchronized hourly measurements of atmospheric and magnetic phenomena for a full year, including at specified dates, measurements every minute and every fifteen seconds, all in the days before photographic or automatic recording or electric or electronic communication, marks one of the high points in the history of scientific co-operation and attention to harmonized measurements.[9]

In our own century, there have been many developments in the scientific field that have expanded on this rich tradition:

– The International Council of Scientific Unions (ICSU) was founded in 1931 to encourage and facilitate international co-operation in scientific activity for the benefit of humankind. One of its specific objectives is to "act as a focus for exchange of ideas, the communication of scientific information and the development of scientific standards, nomenclature units,

PRINCIPLES FOR SCIENTIFIC INVESTIGATION, INTERNATIONAL POLAR YEAR

> "THE EARTH SHOULD BE STUDIED AS A PLANET"
>
> "CO-ORDINATED AND SYNCHRONIZED OBSERVATIONS ARE NECESSARY TO PROVIDE INFORMATION ON CHARACTERISTICS, CHANGES AND THE DISTINCTIVE NATURE OF PHENOMENA IN SPACE AND TIME"
>
> "INTERRUPTED SERIES OF OBSERVATIONS CAN HAVE ONLY RELATIVE VALUE"
>
> "RESULTS OF MUCH GREATER SCIENTIFIC VALUE CAN BE EXPECTED IF STANDARDIZED OBSERVATIONS ARE MADE BY OBSERVERS USING SIMILAR INSTRUCTIONS FOR RECORDING PHENOMENA AT SIMULTANEOUS PERIODS, AND WHO EXCHANGE THE RESULTS OF THEIR OBSERVATONS WITHOUT DISCRIMINATION"
>
> KARL WEYPRECHT 1879

Fig. 2.1.1.3:
Selections from "Principles for Scientific Investigation, International Polar Year"
(Extracted from Heathcote, N. H. de V., and A. Armitage, [1969]. "The First International Polar Year: Origin, Planning and Programme" in Chapman, S., ed., *Anals of the International Geophysical Year,* Vol. 1, Pergamon Press, London, pp. 6–17.)

etc., and in the comparison of methods and intercalibration of instruments"[10]. This work is carried out through the 20 adhering international scientific unions, National Committees of 71 member countries, and numerous specialized scientific committees, among which the Committee on Data for Science and Technology (CODATA) gives particular attention to harmonization of data and data networks.

- The Second International Polar Year, 1932–33, taking advantage of world-wide telegraphic and "wireless" communication technology, helped set in place global networks of "real time" information exchange of observation and measurements of a number of geophysical phenomena.[11] Importantly, also, even during a period of economic depression among the industrialized countries, it helped establish the conviction and the practice that it was in the national interest of many countries to continue to invest public funds in obtaining information about natural phenomena from the world as a whole.

Coming closer to our own time, there has been a continuous proliferation of systems to observe and measure characteristics of our planet. I will mention by name two that are particularly important as background to the Global Observing Systems we are discussing today:

- *The International Geophysical Year (IGY)* 1957–58, which led to the establishment of the World Data Centres, which at first were concerned mainly with collecting and disseminating data, then increasingly have become focused on meeting the needs of "data users".[12]
- *UNESCO scientific programmes,* especially IHP, MAB, IOC and IGCP.

I suspect that we will consider several aspects of the consequences and modern successors to these programmes during this workshop.

At the same time that these scientific programmes have been evolving, developments in the political and policy field began to call for information about the natural environment. The kind of information called for was different from that needed for pure scientific enquiry or resources development. The UN Conference on the Human Environment in Stockholm in 1972 was the first major international expression of this new demand. The need for consistent information about the environment was supported by more than 100 Governments which themselves were beginning to respond to environmental issues and alarms by creating ministries responsible for environmental protection or information.[13] The Stockholm Conference led of course to UNEP, and at the same time to the establishment of the Global Environmental Monitoring System, or GEMS; and from that, to MARC (1975), WCMC and the rest of the modern UN environmental data family.

History of HEM

It may be appropriate to say a few words about the background of HEM, supplementary to the introductory comments of Dr. Baumgardner and Keune, as I may be the only person here today who was directly involved.

In 1984, at the meeting in London of the Economic Summit of leaders of the seven major industrialized countries, a "Group of Environmental Experts" was set up, with the following task:

"Recognizing the role of environmental factors in economic development, to identify specific areas for future research on the causes, effects and means of limiting environmental pollution of air, water and ground and to develop cost-effective techniques to reduce environmental change".[14]

I was a member of this group.

In due course the Group of Environmental Experts reported on actions to be taken with respect to six major areas of concern – atmospheric pollution, toxic and radio-active wastes, marine pollution, pollution of soils and water, inappropriate land husbandry, and climate changes. It also made one general recommendation, and proposed "a study of the improvement and harmonization of techniques and practices of environmental measurement".[15] Two major reasons were given for the need for the study (fig. 2.1.1.4).

One reason was *Scientific*: adoption of internationally consistent technique and practices ensures that research results are comparable; this is particularly important in understanding causes and effects and in monitoring changes.

The other was *Political*: accurate and comparable measurements are vital for the setting and monitoring of environmental standards.

At their meeting in Bonn in 1985, the Economic summit leaders asked the Group to undertake this study.[16] We were assigned tasks which in some respects are similar to the tasks that have brought us to this workshop:

> In environmental measurement, improved anf harmonized techniques and practices are essential:
>
> ■ At a scientific level, the adoption of internationally consistent techniques and practices ensures that research results are comparable; this ist particularly important in understanding causes and effects and in monitoring changes.
>
> ■ At a political level, accurate and compatible measurement is vital for the setting and monitoring of environmental control strategies and standards.
>
> International harmonisation does not mean a uniformity of measurement systems, but rather an increase in consistency of approaches and in transparency of results.
>
> The development of improved measurement techniques and practices should be an ongoing process, not only with respect to procedures already in place, but als as a response to new knowledge and to scientific and regulatory needs.

Fig. 2.1.1.4:
Environmental Measurements: International Comparability and Compatibility of Data
(from Schmidt-Bleek, F., [ed.] [1986] Improvement and Harmonization of Environmental Measurement, GSF Munich, Germany, 12 pp.)

Taking into account the range of environmental issues noted in our report, we were to:

- "consider areas, in all the range of major environmental issues noted in our report, where there is a scientific requirement for improvement and compatibility of techniques and practices, taking into account new and sophisticated technologies";

- "identify areas in which there is cause for concern about the accuracy, precision and sensitivity of physical, chemical or biological measurements of properties of the natural environment";

- "consider and recommend how best to encourage improvements so that those measurements are internationally compatible and recognized as authoritative".

In order to accomplish these, were to consult and liaise with other organizations and programmes involved with environmental data, in particular UNEP, UNESCO and ICSU.

In the course of this work and consultation, we met with many international organizations and discussed their concerns about environmental data inter-compatibility and management. Some of the principal ones were (fig. 2.1.1.5):

This was a wonderful opportunity for members of the Working Group to learn about the different approaches that international organizations were taking to meet the new challenges of obtaining accurate measurements about that dynamic, apparently erratic and little understood subject, the natural environment.[17]

Fig. 2.1.1.5:
Multinational Organizations Concerned with Improvement of Environmental Measurement, 1985–86 (Assembled from Environmental Experts, Economic Summit [1989]. *Report on Ensuring Continuing Progress in Improvement and Harmonization of Techniques and Practices of Environmental Measurement.* GSF Munich, 19 pp.)

Our consultations with other organizations showed, almost without exception, that those responsible for other programmes dealing with environmental data saw a need for a neutral body, not connected with any of the specialized data systems then proliferating, to give attention to the interchange ability of environmental measurements and the compatibility of data describing the environment. It began to appear that a new and different body was needed.

During that period, in 1986, the Committee on Data for Science and Technology (CODATA) of ICSU was faced with the problem of how to apply, to what had been until then largely descriptive information in the biological sciences and assessments of environmental impact, the same standards of rigour and analyses of data that had been evolving in the earth and space sciences. This was required because the International Geosphere-Biosphere Programme, launched in 1984, required interchange of data between the life and geophysical sciences, and there was much interest in developing numerical simulation models that would link biological and geophysical processes. CODATA organized a Workshop Conference on "Directions for Internationally Compatible Environmental Data" which did much to clarify both the scientific needs and scientific limitations with respect to compatibility of data on different environmental subjects under different knowledge systems.[18] I am pleased to recall that Professors Baumgardner and Wiersma were participants in that Workshop, and it is good to see them still in the forefront in this subject. Because of common membership, we were able to achieve some correspondence between the ICSU and Economic Summit activities.

In December 1986 the Group of Environmental Experts submitted its Final Report to the heads of Government of the G-7 Nations of the Economic Summit.[19] The report recommended that the United Nations Environment Programme (UNEP) be asked to institute a body for "Improvement and Harmonization of Technologies and Practices of Environmental Measurements" in co-operation with the International Organization for Standardization (ISO) to provide legal expertise, and the International Council of Scientific Unions (ICSU) to ensure integration with world science programmes and advances. The Group singled out two general priority areas for attention; one scientific and the other political or institutional (fig. 2.1.1.6):

■ AREAS WHERE MEASUREMENTS ARE INADEQUATE BECAUSE OF LACK OF SCIENTIFIC UNDERSTANDING OF THE PROCESSES INVOLVED, OR LACK OF SUITABLE TECHNIQUES.

■ AREAS WHERE TECHNIQUES ARE WELL DEVELOPED, BUT DUE TO PREVIOUS HISTORY, DIFFERING PRIORITIES AND INDEPENDENTLY EVOLVED SYSTEMS, THE DATA PRODUCED ARE NOT COMPATIBLE.

Fig. 2.1.1.6:
Priority Areas for Improvement of Environmental Measurement
(from Environmental Experts, Economic Sumit, [1986]. Final Reports on Improvement and Harmonization of Techniques and practices of Environmental Measurement. GSF Munich, 27 pp.)

I need not tell people in this room that those two types of problems are still with us.

In our report, we identified eight issue areas where, on the basis of our consultations and discussions, international action would be desirable to improve the representativeness and interchange ability of environmental data (fig. 2.1.1.7).

It should be noted that, as our task was to consider areas of direct importance to the economic interests of the Economic Summit, this list focuses on issues of principal concern to technology, economic growth and employment in the industrialized countries. In our report we assumed, in a way that seems today to neglect many other problems, that attention to these issues would benefit the whole world.

It is useful to note also, that we made a conscious attempt not to assess the adequacy of existing data, but focused on whether or not the measurements made to produce the data were (i) scientifically compatible, and (ii) being produced in a way that would allow others to understand them, compare them, and repeat them.

(I) KEY INDICATORS OF ENVIRONMENTAL QUALITY AND EARLY DETECTION OF ENVIRONMENTAL CHANGE;

(II) SOURCES, DISTRIBUTIONS, VARIATIONS, AND EFFECTS OF:

    A) STRATOSPHERIC OZONE,

    B) OZONE AND OTHER PHOTOOXIDANTS IN THE TROPOSPHERE;

(III) LONG-RANGE TRANSPORT AND EFFECTS OF AIR POLLUTANTS, INCLUDING ECOLOGICAL EFFECTS OF ACID DEPOSITION;

(IV) ENVIRONMENTAL BEHAVIOUR AND EFFECTS OF PERSISTENT CHEMICALS;

(V) BEHAVIOUR OF POLLUTANTS IN SOILS AND SEDIMENTS;

(VI) PREPARATION AND SUPPLY OF REFERENCE MATERIALS AND LONG-TERM ARCHIVING OF CHARACTERIZED ENVIRONMENTAL SAMPLES;

(VII) MEASUREMENT OF KEY PARAMETERS INVOLVED IN GLOBAL CLIMATIC CHANGE AND OF ITS CONSEQUENCES;

(VIII) IDENTIFICATION, BEHAVIOUR, AND EFFECTS OF KEY SPECIES OF POLLUTANTS IN NATURAL WATERS, AND THEIR CONSEQUENCES AS REGARDS CONSERVATION AND PROTECTION OF NATURAL RESOURCES.

Fig. 2.1.1.7:
Major Environmental Issues Requiring International Improvement and Harmonization of Environmental Measurement
(from Environmental Experts, Economic Summit, [1986]. *Final Reports on Improvement and Harmonization of Techniques and Practices of Environmental Measurement.* GSF Munich, 27 pp.)

The report of the Group of Environmental Experts was accepted by the G-7 Heads of Government at their 1987 meeting in Venice, and the recommendation was passed to the UNEP Governing Council, which in turn endorsed it. The German Government offered to host a small office, and after some bureaucratic delays, HEM came into existence in 1989.

Where are we today?

Quite a lot has changed since 1989, and most of those changes are accelerating even as we meet today. We can agree, quite easily I think, on the sub-title for our Workshop "To develop a consistent view of the environment world-wide". At the same time, most of us also probably feel that if such a view is to be useful to those who have responsibility for developing policies that will result in changes in institutional practices and societal behaviour that can avert the deterioration of the planet, we have to try to understand the way data systems themselves are changing, quite aside from the environmental changes that the data purport to show. We will have to "harmonize" data and data systems in a way that will be most useful for tomorrow, not for yesterday. The changes in data systems that seem to be important are of four kinds: changes in the type, technology, capacity and management of environmental data systems themselves; changes in the environment-related issues for which environmental data are needed; changes in scientific understanding about planet Earth, environmental processes and human behaviour; and changes in political and societal structure and decision-making.

Changes in data technologies and systems

I need not tell anyone in this room of the enormous advances in the technical data collection made in the last decade, to produce observations and measurements of all kinds and at all scales in a quantity and with an accuracy that could scarcely be dreamed of when the Group of Environmental Experts first began to discuss the need for harmonization. Nor do I need to dwell on the revolution in data management and display. But it is necessary to point out, I think, that many of the advances in environmental observation and data management are towards technical sophistication, which favours only parts of the world and its institutions. Many countries and large parts of humankind which desperately need better environmental information cannot afford, or make use of data from remote sensing, automatic recording or auto-analysis devices or get much out of a CD-ROM disc or a Geographic Information System; and although satellite surveillance can obtain data from every part of the planet, the data obtained are likely to be held by, and easily accessible to only a few. Are we embarking on a new kind of scientific colonialism? In many parts of the world, there is a growing real resentment that rich and technically advanced countries have information about less-advanced countries that those countries themselves do not have.

These issues raise different and serious issues of harmonization: are the data to be harmonized within the club, as it were, for the technically sophisticated only; or is a "consistent view of the environment world-wide" to work toward providing compatible information from all sources, ranging from satellite observations to the valuable and irreplaceable seasonal environmental knowledge of the farmer in the rice paddies? Do we (or does someone) have an obligation to produce the information in a form that the rice farmer can use? We should not be too quick to answer that question, and should think about the implications of our answer.

Along with the technical changes in data collection and management, there has of course been an enormous proliferation of national and international data banks and data systems, each necessary and justified for its purpose or subject but with various needs for, or susceptibility to, harmonization between their respective contents. HEM has made a valuable start at addressing this question, through its red book, blue book and green books. And this morning I saw for the first time the HEM yellow book. But it is an uphill fight, to achieve compatibility between entrenched independent systems each of which is fighting for its own survival. The first steps are to produce directories of who are in the field and what are they doing. Then comes the harder job of compiling what are they producing, and presenting the mismatches and inconsistencies in such a way that those responsible will see the advantages, to them, of greater compatibility. These are areas where IGBP presents a challenge to the World Data Centres and IGBP-DIS; and where, I hope, GTOS and HEM will develop a mutually beneficial partnership.

Changes in environment-related issues for which data are needed

The topics of the major environment-related issues of global or regional concern have not changed much in the past twenty years. They were identified in broad outline at the UN Stockholm conference, repeated as best we could for the industrial countries and tied to economic performance in our 1987 report to the Economic Summit where the need for harmonization of measurements was noted, re-phrased for the world at large in the World Conservation Strategy,[20] and of course they were outlined by WCED[21] and were the basis for the UNCED at Rio de Janeiro in 1992. The main change is that these issues have nearly all become much more serious and more urgent, and the need for reliable and compatible information is much greater. This increased seriousness in turn puts urgency and importance to harmonization of techniques and practices of measurement. Another related change is that these issues are no longer recognized only by scientists and the environmental community, but there is now widespread political and popular awareness of the seriousness of the environmental situation. There is a growing demand at all levels in both developed and developing countries to be informed about what is happening to the planet, to the environment, and to ourselves as human individuals and societies.[22] This demand can be met only if observations and data from different sources on unlike subjects can be brought into line and amalgamated – i.e. harmonized – to give reliable descriptions or assessments of the environmental situation at any one place or time compared with other places or with the past.

Changes in scientific understanding

Scientific knowledge and understanding often seem to advance in spurts, between periods of relative solidification or much slower progress. Recently, there has been no doubt that we are in a period of rapid advance in our knowledge of planetary and biological processes and of the human animal and its behaviour. One need hardly look further than the IGBP or Global Change programme, with its ambitious objective that would have been considered preposterous twenty-five years ago but which now is a matter of organizing, financing and managing co-ordinated research operations, assembling compatible data, and presenting scientific questions in ways that are seen to be relevant to national priorities and socio-economic issues. It is also, of course, a matter of achieving intellectual co-operation between individualistic researchers! The objective, you will recall, is nothing less than

„To describe and understand the interactive physical, chemical and biological processes that regulate the total Earth System, the unique environment that it provides for life, the changes that are occurring in the system, and the manner in which they are influenced by human actions."[23]

Obviously, compatible and harmonized observations and measurements in a global perspective are basic to the Global Change programme. Many of the core programmes and regional studies of IGBP are very much involved with data systems and harmonization of measurements. The IGBP Data and Information Systems Office (IGBP-DIS) has a data co-ordinating responsibility. But it is important to note that the IGBP objective does not mention data or information; its purpose is to describe and understand. Its defined area of action is four or five steps up the "staircase of knowing".

Several of the important international scientific activities, in addition to the IGBP, have identified current projects dealing with compatibility of data. Some are listed in Figure 2.1.1.8.

I will not describe these individually, but draw your attention to the fact that while these various studies and programmes are making progress in achieving *internal consistency* of their data, they cannot be expected to develop *inter-programme overall compatibility*. That is a task for CODATA and for HEM.

ICSU	–	International Council of Scientific Unions		
		ICL	–	Inter-Union Commission on the Lithosphere
		IGBP	–	International Geosphere-Biosphere Programme
		SCOPE	–	Scientific Committee on Problems of the Environment
		SCOR	–	Scientific Committee on Oceanic Research
UNESCO	–	United Nations Educational, Scientific and Cultural Organization		
		MAB	–	Man and the Biosphere Programme
			–	Biosphere Reserves
			–	Ecosystem Change Under Human Impact
		IHP	–	International Hydrological Programme
WMO	–	World Meteorological Organization		
		WCRP	–	World Climate Research Programme

Fig. 2.1.1.8:
Major International Scientific Programmes Undertaking Improved Measurement of the Environment
(Original Figure – to illustrate text by E. F. Roots.)

Refreshing features of the current surge of scientific knowledge about the environment are the way that studies of natural phenomena – climate change, for example, or environmental influences on genetic evolution – are seen to be relevant to the human condition;[24] and the new, almost revolutionary openness to different sources of knowledge, with admissions that some areas are presently "unknowable" or beyond the reach of disciplinary research.[25] Both of these developments present new challenges to the harmonization of data. The former was inevitable as the advance of science revealed the interactions between humans and the natural world in a new light, and as worsening environmental problems have brought science into the field of policy and citizen concern. The latter has come from a growing awareness that the traditional knowledge of many indigenous societies contains environmental understanding and wisdom that although not accessible to established disciplines of research, is relevant to current serious environmental issues,[26] and from a sober realization that inherent processes of symbiosis and autopoesis in the biosphere are only poorly glimpsed by science so far, although their effects on a planetary scale are evident.[27] Along with its new advances in knowledge, science in many areas is learning a new humility. Some of that humility must be a characteristic of our approach to harmonization.

Changes in decision-making

The changes in scientific understanding and perspective, the increasing seriousness of large-scale environmental problems, and the advances in data capability and supply have all contributed to a broadening and diversity of what the so-called "user" wants or can make use of from environmental data. These changes also affect the forms in which the resulting information should be supplied, to be effective. At the same time, the fundamental changes in the structures of modern societies in both industrially-developed and developing countries, and the confusing intermingling of societal elements and different coercion/cognition/production relationships, noted earlier, have resulted in demands for environmental information, which may be used by different interests for quite different purposes. The same environmental and resources information is seen to be essential for policy development (coercion), for improved economic production or performance, or as a basis for cognition as influence in its own right. These diversities and tensions were evident at the UNCED in Rio, and are reflected in the Rio Declaration, Agenda 21, the various conventions and protocols that were signed or put aside, and in the several critical statements and analyses from public interest groups. All of these documents and statements used amalgamated environmental data to describe the global conditions, but they used it for their various respective purposes. To what degree the data used were or could be harmonized to achieve a semblance of compatibility is another matter. In some cases the internal consistency appears to have been quite good; in others the numbers were, at best, chalk and cheese side by side.

The issues aired at UNCED and brought together by UNEP and UNESCO in follow-up have handed the scientists of the world a new series of tasks regarding environmental measurement. Few of the issues are wholly new; it has been the bringing together of research and data collection with an openly expressed political, social and economic urgency that has posed, more clearly than ever before, such questions as:

– What kinds of decisions must be made, and by whom, to affect human actions in order to arrest present environmental and resource deterioration?
– What kinds of information are needed if the "right" decisions are to be made and their effects assessed?
– How can the need for information be translated into effective and sustained data-gathering and environmental measurement?
– What kinds of harmonizations are needed or desirable, to meet the needs of major policies and monitoring that require amalgamated data, and yet need to preserve the individualities and anomalies that may be more important?

A major response, at the international institutional level, has been the conception of three linked Global Observing Programmes, GTOS, GCOS and GOOS (fig. 2.1.1.9).

We will spend most of the next three days exploring issues of design of GTOS and interactions with the climate and ocean observing systems to produce harmonized data. The six sponsoring international bodies have their own priorities and needs for consistent data.

Note that UNEP, like WMO, is involved in all three systems. HEM will or should be likewise involved in all, and more importantly, in facilitating overall harmonization.

The Human Dimension of GTOS – Larger Problems

The Global Observing Systems, GTOS in particular, have because of their parentage in UNEP, UNESCO and UNCED concerns, a different dimension than almost all world-wide scientific and environmental observing and monitoring systems carried out until now. For they will be observing not only the effects of human activities on the environment (IGBP does that also), but will as an integral part of the same system be gathering data on humans themselves as active components of the global ecosystem. It seems to me that in many ways the scientific community responsible for designing and operating the Global Observing Systems has a philosophical problem in this respect that it has ignored or been reluctant to face.

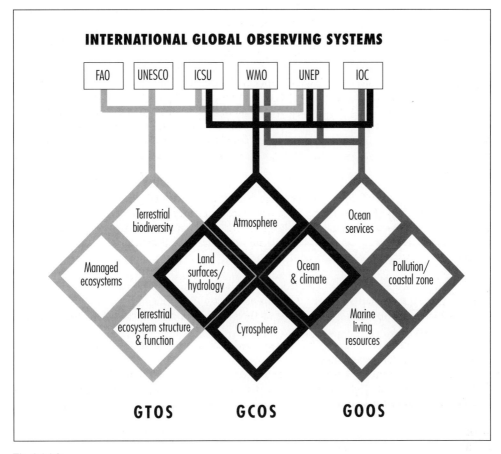

Fig. 2.1.1.9:
Relationships GTOS, GCOS, GOOS
(from Heal, W. O., Menaud, J.-C. and Steffen, W. L. [eds.] [1993]. "Towards a Global Terrestrial Observing System [GTOS]", *MAB Digest 14/IGBP Global Change Report 26,* UNESCO, Paris, 71 pp.)

It has been said been said repeatedly, e.g. in the reports of GTOS planning meetings[28, 29] that the observing systems must "include experiments, both managed and unmanaged systems", must include observations on "human populations and settlements and their impacts on the land and biota", and that the information is required by "users", "to advise governments on the state of the global environment, for regional and global early warnings, for regional action plans, to determine development priorities", and so on. In other words, in GTOS we are planning to establish a system to observe ourselves and everybody else, all of us together, all over the globe, to produce information that then will be used by some or many of us, in little bits and pieces of countries and societies that have different value systems and different sensitivities and different ways of controlling or encouraging human behaviour or making use of environment. There will undoubtedly be quite different opportunities, options and convictions as to what we can or should do in light of the information we will obtain.

Despite the logic and the easily explained need for a comprehensive harmonized global environmental observing system, the fact that such a system must necessarily and by design include socio-economic, demographic, and human-influences data means on the one hand that GTOS will include data on human values and coercion mechanisms and their effects (i.e. will be monitoring the effects of policies), and on the other hand that GTOS itself will become an instrument in the changing structures of society. On the information autobahn to the future, the Global Observing Systems will, if I can push the analogy perhaps too far, be both roadbed and vehicle.

It seems to me that we all have a responsibility to think through carefully the implications of these human-society-related aspects of successful Global Observing Systems. Will the global data widen the gap between the information-rich and the information-poor on this planet? Do the Global Observing Systems, by making comprehensive information not only about our shared environment but about each and all of us and our country available mainly to the technically sophisticated, bring serious issues of *human rights,* collective rights as well as individual rights, and rights to a healthy environment?[30] Is the managed and industrial environment, affected and controlled by private enterprise or government policy, also part of the global commons to be observed at will by anyone with the technology to do so? Up until now, these questions have arisen mainly in connection with military or commercial proprietary issues. But with global observing systems that gather and harmonize data on people and their activities (as they must, for people and their activities are major forces driving environmental change), as a "basis for planning and executing appropriate land and water management for sustainable development and national action plans" (GEMS Report 24) questions of sovereignty, freedom of action, equity, and opportunity to capitalize on perceived impending needs or weaknesses are not far away.

It is easy for those of us preoccupied with the scientific and technical issues, and who are already philosophically attuned to wanting to preserve a healthy environment for the future, to persuade ourselves that more and better environmental information from all parts of the planet will make it more possible for humankind to develop policies and behavioural patterns that are more sustainable than the ones that have brought our world to its present state. But our experience with the information provided by sub-global or regional observing systems has not been reassuring. Will the global information autobahn be safer and have a better destination than the footpath of country knowledge?

None of these reflections are in any way to suggest that Global Observing Systems, GTOS in particular, are not needed, or that the information they deliver should not be a force for good to the world as a whole. But the Global Observing Systems are a tool, not a solution to our problems. The tool can be used to make our problems worse, or to help us solve them. It is what we use the tool for, not how technically good it is or how skilfully it is used, that will determine the human and planetary result. Those of us who are busy in designing the tool and putting it into operation should bear in mind that history shows that improvements in technology and information rarely have the simple beneficial effects expected.

Because of these issues, bearing in mind the "global problematique" that has brought the five sponsoring agencies together to support a GTOS, it may not be wise to look on the Global Terrestrial Observing System simply as a co-ordinated multi subject extension of systems presently in place under UNEP, WMO, ICSU, etc. The world-wide observation and data-gathering programmes in meteorology, geophysics, tides and sea level, etc. primarily serve the technical needs of a major user or client, even though all citizens may be affected by the phenomena observed. The support for such observing systems, while not always adequate, has been relatively straightforward because the weather forecaster, or the person who needs to determine the location and magnitude of an earthquake has a direct need for world-wide data. When these specialized observing systems begin to serve a broader set of interests, as is happening in recent years with the weather services, difficult issues of continued support, "user pay", responsibility for interpretation, etc. arise quickly. GTOS has by intention a much wider clientele and is expected to have a much more diffuse range of "users" than any of the observing or monitory systems now in place.

In some ways, perhaps, the forerunner of GTOS should be not the present international technical observing programmes or even the British Admiralty Manual, but the country-wide survey that led to the Domesday Book in England in 1086. After he conquered Anglo-Saxon England and expropriated the lands of the entire country, William of Normandy sent commissioners to every shire and county to gather data on "the lands, the riches and appurtenances of the realm", to determine "the holdings of the King, the churchmen, of the barons, of the women, the number of freemen, the tenants, the meadows and farms, oxen and pigs, plowshares and mills".[31] This was more than seven hundred years before any census or systematic geographical survey was established. The results of this extensive multi-discipline survey were compiled into one comprehensive definitive information base, the Domesday Book. The purpose, clearly, was to provide a tax base and set up a national administrative control system; to provide a harmonized information base for what we would today call new policy development and development planning.

Historians may still debate about the effect of the Domesday Book on the subsequent history of England. But within five years there was established in all parts of the country standardized but separate judicial and financial administration machinery (the officials of the latter carrying a checkered table-cloth to each county seat, on which illiterate barons could compute their revenues and payments – hence "exchequer" – the earliest digitized geographic information system?), all operating on a common and openly accessible information base. There seems little doubt that this single comprehensive survey and harmonized information system gave England a jump start over its European neighbours in becoming a unified nation, despite its heterogeneous composition and internal tensions, with co-ordinated but separate legal, trade and educational systems. Can GTOS help to do something similar, on a world scale, nine hundred years later?

Harmonization and Back to Nature

The global observing systems do not and must not deal only with large-scale data. It is equally important that attention be given to harmonizing the basic elements of measurement of characteristics and processes across a range of scales, from a single cell to changes of continental or global dimensions (fig. 2.1.1.10).

Clearly, different measuring and sampling techniques and data handling systems are needed at each scale.[32] But a consistent approach and relationship must be developed so that the information can be "nested" in a harmonious way up and down the scale.

Harmonization of measurements over a *range of time* scales is even more difficult, because observing methods change with time or become inappropriate at different process rates. In contrast to spatial scaling, where it is possible to move to a large scale by making observations

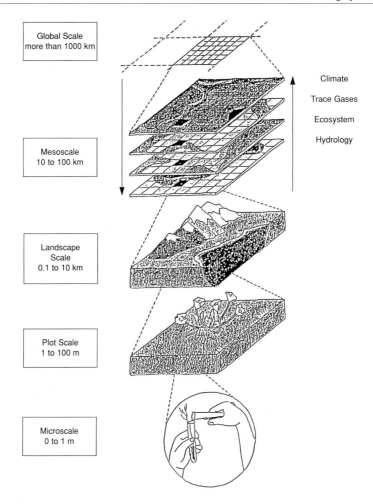

Fig. 2.1.1.10:
A Hierarchy of Spatial Scales
(from McCauley, L. l. and M. F. Meier, [eds.] [1991]. Arctic System Science: Land/Atmosphere/Ice Interactions. A Plan for Action. Arctic Research Consortium of the U.S., Fairbanks, Alaska, 48 pp.)

over a larger area, there is no way that one can extend the record to a longer time period if the observations have not been taken. Proxy data may help, but they remain "proxy". But harmonization of data from different time scales is essential if the dynamic nature of the environment is to be documented, and the changes due to human activities or remedial actions assessed.

Any broad assessment of environmental quality must measure a number of unlike things and reduce the results to more common format or scheme for comparison. Figure 2.1.1.11 is borrowed from one of Lars-Otto Reiersen's publications of the Arctic Monitoring and Assessment Programme[33] and shows the large number of environmental media that must be sampled, each with its appropriate technique, to give a harmonized picture of heavy metals in the arctic environment.

Program	Media	Metals												
		Cd	Cu	Hg	Pb	Zn	Cr	Ni	As	Se	Ai	V	Fe	Li
Atmospheric	AIR	E	ES	E	E	ES	ES	ES	ES	ES	ES			
	Precipitation	ES	ES	ES	ES	ES	ES	ES	ES	ES	ES			
	Snowpack	R	R	R	R	R	R	R	R	R	R			
Terrestrial	Soil	E	E	ES	E	E	E	E	E	E	ES	E	E	
	Humus	E	E	E	E	E	E	E	E	E	E	E	E	
	Mushrooms	R	R	R	R	R	R	R	R	R	R	R	R	
	Lichens	R*	R*	R*	R*	R*	R*	R*	R*	R*	R*	R*	R*	
	Mosses	E	E	ES	E	E	E	E	E	E	ES	E	E	
	Higher plants	R	R	R	R	R	R	R	R	R	R	R	R	
	Mammals /Birds	E	E	E	E	E	R	R	E	E	E	R	R	
Freshwater	Sediment	E	E	E	E	E	R	E	E	E	R		R	
	Water	E	E	R	E	E	E		E	E	R		E	
	Susp. solids[1]	E	E	E	E	E	E	E	E	E	R		E	
	Biota	R	R	E	R	R	R	R	E					
Marine	Sediment	E	E	E	E	E	R	R	R		E			E
	Water	R	R	R	R	R	R	R	R					
	Susp. solids[1]	R	R	R	R	R	R	R	R					
	Biota	E	R	E	E	R	R	R	R	E				
Human Health	Blood	E	E	E	E	E		ES		E				
	Tissue													
	Placenta													
	Urine	R						ES	R					
	Hair			R										
	Breast milk													

Fig. 2.1.1.11:
Monitoring Metals in the Arctic
(E: essential, ES: essential sub-regional, R: recommended, *): essential as a part of food web studies, otherwise R)
(from Reiersen, L.-O., [1992]. Arctic Monitoring and Assessment Programme, Phase One, Technical Report. Arctic Monitoring and Assessment Programme Secretariat, Norwegian State Pollution Control Authority [SFT], Oslo, Norway)

To harmonize environmental data properly, we must not get so involved in the mechanisms and techniques and numerical management that we lose sight of the fact that it is *Nature* that we are sampling and measuring.

Figure 2.1.1.12 is a simplified cartoon of an Arctic landscape, with some major environmental interactions identified (solid arrows).[34] These can each be measured individually, although to get representative and consistent measurements poses severe technical problems and is a difficult and expensive task. The changes that are occurring to this landscape right now are shown by the open arrows. These are much more selective, in some cases site-specific but no less important on a regional or global scale, and they pose quite different problems of harmonization, even through the actual physical or chemical measurement may be the same as for the "undisturbed" data. Both sets of measurements, compatible (harmonized?) at appropriate scales are vital to a successful GTOS, GCOS, GOOS.

There is, finally, the problem of how harmonized data gets fed into the decision process without losing sight of the fact that it is the idiosyncrasies of the natural world that we are measuring. Let me give an illustration from the Arctic, which I know best.

Figure 2.1.1.13 is a sketch of an Arctic scene – snow covered land, sea ice, open ocean with icebergs, whales and seals, all natural except that somewhere over the horizon factories and power plants are spewing out pollution. The picture is harmonious if not harmonized.

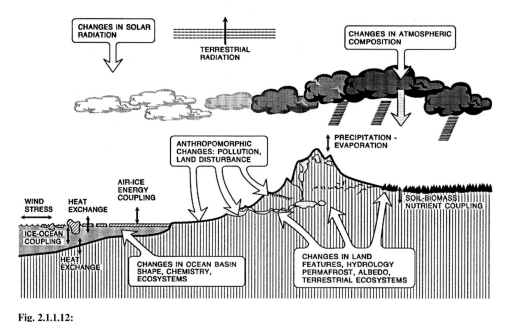

Fig. 2.1.1.12:
Landscape Interactions and Changes (Arctic)
(Adapted from Eddy, J. A., M. F. Meier, and E. F. Roots, [eds.] [1988]. *Arctic Interactions*. Office for Interdisciplinary Earth Studies, Boulder, Colorado, 45 pp.)

He or she sets up a weather station. Ships probe the ocean. Aircraft and satellites measure and record everything in sight. Somebody samples the sediments, measures the ice, counts the plankton and the seals. Each gets his or her little bits of measurement and data, according to his or her techniques or interests or the "national priorities". Now the picture is not harmonious at all. Instead of a landscape we have a heterogeneous assemblage of measurements, observations and data (fig. 2.1.1.15a).

These bits of information are likely filed away in separate data sets.
(if we take this picture away from its background and look at it as measurements and data, it is even harder to make sense of it (fig. 2.1.1.15a). So, now come the magic synthesizers, the modellers and the computer experts (fig. 2.1.1.16).

They can make sense of all this. Put the data and the models into a computer, find a programme that relates all this to some priority issues such as how did the seals get contaminated with mercury, and Presto! — we have status reports and environmental baseline or disturbance maps. The data are once again harmonized. *Or are they?* They are however, now the basis for decisions.

But the decision maker, once given the question on which a decision must be made, has to put those data and the synthesized information through a series of filters

- What political authority is there for decision at that level?
- What are the economic, social, etc. consequences of the alternative actions and how does the boss, or the public value *them?*
- What is just *common* sense?

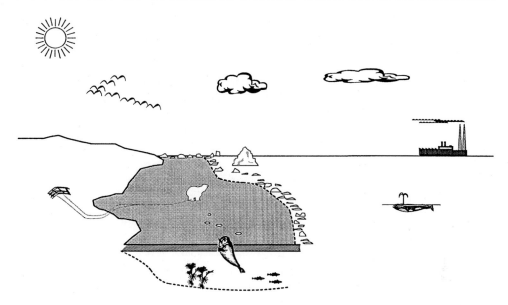

Fig. 2.1.1.13:
Arctic Scene
(Figure to illustrate text.)

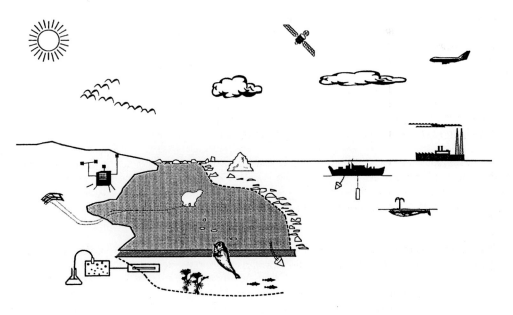

Fig. 2.1.1.14:
Scientific Observation in the Arctic
(Figure to illustrate text.)

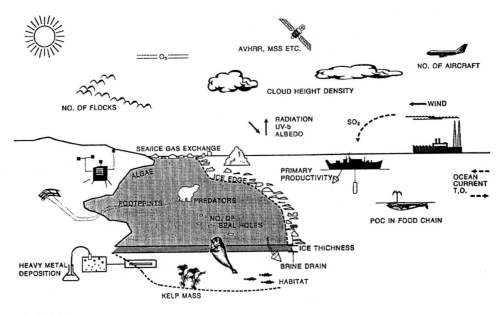

Fig. 2.1.1.15a:
Data and Information from Scientific Observation in the Arctic
(Figure to illustrate text.)

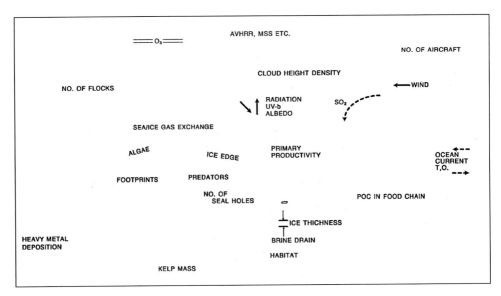

Fig. 2.1.1.15b:
Arcitc Datas and Information
(Figure to illustrate text.)

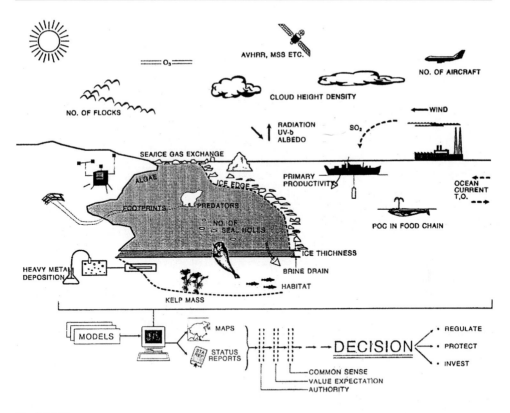

Fig. 2.1.1.16:
Decision based an Arctic Observations
(Figure to illustrate text.)

These filters are subjective; and yet they are perhaps the best hope of bringing harmonization of disparate measurements back to Nature.

As a final thought, we should not pretend that even the best harmonized data, from the best possible observing system will give a full picture of the total environment. It will only tell us about parts of it. We should remember the words of the Danish mathematician and poet Piet Hein, who likened the study of natural science to solving a cross-word puzzle:[35]

> *The world is a cross-word, involved and immense*
> *With a datum that fits in each spot;*
> *And the teeniest, tiniest details make sense,*
> *But the overall pattern does not.*

Notes and References:

1. GELLNER, E., (1991). *Plough, Sword and Book, The Structure of Human History*, Paladin Grafton Books, London.
2. ROOTS, E.F., (in press). "Ethics, Environment and Sustainable Development Policies", in *An Environmental Ethics Perspective on Canadian Public Policy*, P. Crabbé. ed., University of Ottawa Press, Ottawa.
3. HEIDEGGER, M. (translated by J. M. Anderson and E. H. Freund), (1966). *Discourse on Thinking*, Harper and Row Publishers, New York.
4. There are many examples of this change, which in turn has penetrated the agrarian and industrial societal statements, if not their practices. See for example the *Rio Declaration on Environment and Development* (United National General Assembly, (1992); or *"Caring of the Earth; a Strategy for Sustainable Living"*, issued by IUCN/UNEP/WWF, (1991). A more local example familiar to the author is the Government of Canada publication *"Canada's Green Plan for a Healthy Environment"*, Ottawa 1990.
5. ROOTS, E. F., (1992). "Environmental Information – A Step to Knowledge and Understanding." *Environmental Monitoring and Assessment*, No. 20, pp. 87–95.
6. CAWOOD, J., (1977). "Terrestrial Magnetism and the Development of International Collaboration in the Early Nineteenth Century", *Annals of Science*, 34, pp. 551–587.
7. HERSCHEL, J. W. F., ed., (1851). *Admiralty Manual of Scientific Enquiry: Prepared for the Use of Officers in Her Majesty's Navy; and Travellers in General"*, British Admiralty, London. The contributing authors included many of the best-known scientists of the nineteenth century – e.g. Airy, Beaufort, Darwin, Euler, Hooker, Richardson, Sabine and others.
8. HEATHCOTE, N. H. de V., and A. ARMITAGE, (1959). "The First International Polar Year: Origin, Planning and Programme", in Chapman, S., ed., Annals of the International Geophysical Year, Vol. 1, Pergamon Press, London. pp. 6–17.
9. ROOTS, E. F., (1982). "Anniversaries of Arctic Investigation: Some Background and Consequences", *Trans. Royal Society of Canada*, Series IV, Vol. XX, pp. 373–390.
10. ICSU Yearbook (Issued Annually), ICSU Press. Paris.
11. BARTELS, J., and others, (1959). "The Second International Polar Year". *Annals of the International Geophysical Year*, Vol. I., Pergamon Press, London, pp. 205–382.
12. SEE, for example, Shapley, A. H. et. al.,(1987). *Guide to the World Data Centre System*, issued by the Secretariat of the ICSU Panel on World Data Centres, Boulder, Co., U.S.A., pp. 1–9.
13. WARD, B. and R. DUBOS, (1972). *Only One Earth: the Care and Maintenance of a Small Planet* (Commissioned by the U.N. Conference on the Human Environment). W. W. Norton, New York, 225 p.
14. *Report on the Environment* by the Technology, Growth and Environment Working Group (to the Economic Summit) (1985). Her Majesty's Stationery Office, London (U.K.).
15. Ref. (14), Appendix A, p. 13.
16. Technology, Growth and Employment Working Group, (1985). *Report on Further Steps Regarding Improvement and Harmonization of Techniques and Practices of Environmental Measurement*, Federal Minister of Research and Technology, Bonn.
17. Environmental Experts, Economic Summit, (1986). *Report on Current International Scientific Activities in Improvement and Harmonization of Techniques and Practices of Environmental Measurement*, GSF Munich.
18. ICSU-CODATA, (1986), (1988). *Directions for Internationally Compatible Environmental Data – A CODATA Workshop*. ICSU Press, Paris (Position Papers, 1986; Report 1988).
19. Environmental Experts, Economic Summit, (1986). *Final Report on Improvement and Harmonization of Techniques and Practices of Environmental Measurement*, GSF Munich, p. 27.
20. IUCN/UNEP/WWF, (1980). World Conservation Strategy: Living Resource Conservation for Sustainable Development. International Union for Conservation of Nature and Natural Resources, United Nations Environment Programme and World Wildlife Fund, Gland, Switzerland.
21. World Commission on Environment and Development (WCED), 1987. Our Common Future, Oxford University Press, Oxford.
22. UNESCO, (1992). Declaration of Belem. UNESCO International Forum on Science and Culture for the 21st Century: Towards Eco-Ethics, Alternative Visions of Culture, Science, Technology and Nature, Belem, Brazil. UNESCO Press, Paris.
23. The International Geosphere-Biosphere Programme: A Study of Global Change, (1994). *IGBP in Action: Work Plan 1994–1998*. IGBP Global Change Report No. 28. IGBP Secretariat, Stockholm.
24. See, for example: (i) Mungall, C., and D.J. McLaren. eds., (1990). *Planet Under Stress: The Challenge of Global Change*. Oxford University Press. Toronto; and (ii) *For Earth's Sake; A Report from the Commission on Developing Countries and Global Change*, (1992), International Development Research Centre, Ottawa; and (iii) *Our Planet, Our Health*, (1992), Report of the WHO Commission on Health and the Environment, World

Health Organization, Geneva.
25 Many recent papers and symposia, often focused primarily on themes of environmental sustainability or the frontiers of scientific progress, contribute increasingly to this theme. Some that explore the limits and role of scientific information include: (i) UNESCO (1990). *Vancouver Declaration,* Final Report of the UNESCO Symposium on Science and Culture in the 21st Century, Canadian Commission for UNESCO, Ottawa (especially papers therein by Akyeampong, Riman, Nakamura, Elmandjra); (ii) Bauer, H. H., (1992). *Scientific Literacy and the Myth of the Scientific Method,* University of Illinois Press, Urbana; (iii) Fetzer, J. H. (ed.), (1993). *Foundations of Philosophy of Science:* Recent Developments, Paragon Issues in Philosophy, Paragon House, New York.
26 For example: Brooke, L., (1992). Science and the Experiences of the Inuit of Nunavik, Arctic Monitoring and Assessment Task Force, Toronto. Report issued by AMAP Secretariat, Oslo. Also UN International Indigenous Commission, (1994), "A Report prepared for the United Nations Conference on Environment and Development". See also Principle 22 of the Rio Declaration on Environment and Development, and Chapter 26 of "Agenda 21".
27 For example, the controversies stirred among a variety of leading scientists by the writings of J. Lovelock and L. Marqulis, e.g. Mann, C., (1991), in *Science,* vol. 252, pp. 378–381.
28 HEAL, O. W. et. al. (eds.), (1993). *Towards a Global Terrestrial Observing System (GTOS).* MAB Digest 14/ IGBP Global Change Report 26, UNESCO, Paris.
29 UNEP, (1994). *Report of the First Meeting of the Ad Hoc Scientific and Technical Planning Group for a Global Terrestrial Observing System (GTOS).* UNEP/GEMS Report Series 24, Geneva.
30 ROOTS, E. F., (1993). "The Right to Environment: Population, Carrying Capacity and Environmental Processes", in Mahoney, K. E. and P. Mahoney, (eds.), *Human Rights in the Twenty-First Century, A Global Challenge.* Martinus Nijhoff Publishers, Dordrecht., pp. 529–549.
31 GALBRAITH, V.H., (1961). *The Making of Domesday Book.* Oxford University Press, Oxford.
32 Many technical papers have been written on these problems. A simple non-technical presentation is in McCauley, L. L. and M. F. Meier, eds., (1991). *Arctic System Science: – Land/Atmosphere/Ice Interactions, a Plan for Action,* Arctic Research Consortium of the U.S., Fairbanks, Alaska, Chapter 3, "An Approach to Understanding and Predicting Global Change in the Arctic", from which Figure 10 is taken.
33 REIERSEN, L.-O., (1992). *Arctic Monitoring and Assessment Program, Phase One.* Technical Report. AMAP Secretariat, Oslo.
34 Adapted from Eddy, J. A., M. F., Meier and E. F. Roots, eds., (1988). *Arctic Interactions.* Office for Interdisciplinary Earth Studies, Boulder, Colorado, p. 3.
35 HEIN, P., (1984). Grooks VII, Borgen's Pocketbooks 174, Copenhagen.

2.1.2 Earthwatch: An introduction to the UN system-wide coordination of environmental observation

A. L. Dahl

Earthwatch was the name given by the 1972 United Nations Conference on the Human Environment in Stockholm environmental assessment part of its action plan, covering the whole UN system and catalyzed/coordinated by UNEP. It has included the Global Environment Monitoring System (GEMS), the Global Resources Information Database (GRID), a global referral service (INFOTERRA), the International Register of Potentially Toxic Chemicals (IRPTC), UNEP's State of the Environment Reports, and a numbers of cooperative activities in the UN system, but it has never had the resources to fulfil its broad mandate to assess and report the global environment. An early Inter-agency Working Group on Earthwatch faded out in 1980.

In 1989, the UN General Assembly recognized the need to strengthen Earthwatch and called for broader participation in its as a continuing mechanism "to make authoritative assessments, anticipate environmental degradation and issue early warnings to the international community". Again in 1993 in a resolution on environmental monitoring, it drew attention to the importance of participation by all relevant parts of the United Nations system in Earthwatch, in particular in its environmental monitoring programmes, and the need for early warning capabilities in those programmes, and emphasized "the need to make Earthwatch a more efficient instrument for environmental sensing and assessment of all elements influencing the global environment in order to ensure a balanced approach in serving, in particular, the needs of developing countries". It requested a report on environmental monitoring, containing proposals and recommendations within the context of Agenda 21 and a review of Earthwatch, to be prepared in 1994 by UNEP in cooperation with relevant entities within and outside the United Nations system.

Agenda 21 adopted the Rio "Earth Summit" in 1992 calls for strengthening Earthwatch. It included among the priority areas for UNEP: "environmental monitoring and assessment; both through improved participation by the United Nation system agencies in the Earthwatch programme and expanded relations with private scientific and non-governmental research institutes; strengthening and making operational its early warning function". Agenda 21 notes that Earthwatch has been an essential element for environment-related data, and calls for an equivalent and complementary "Development Watch" which should then be coordinated with Earthwatch to ensure the full integration of environment and development concerns to meet the information and reporting requirements of the Commission on Sustainable Development.

In general terms, Earthwatch is a mechanism to stimulate the UN system, in collaboration with governments and the international scientific community, to gather, integrate and analyze data and information as the basis for comprehensive assessments of environmental issues, early warning of environmental threats and the identification by the international community of policy options and management responses for achieving sustainability. The role of UNEP, as the principal body in the UN system in the field of environment, is to catalyze the Earthwatch process, to build increasing results through Earthwatch, and to coordinate reporting on Earthwatch to the Commission on Sustainable Development.

The central aim of Earthwatch should be to make the provision of environmental information for decision-making a more coherent and effective process as called for in Agenda 21. This requires the integration of the many kinds of information required to achieve sustainable development, relating scientific data on the environment to issues such as poverty, types of development, resource consumption and quality of life, with an emphasis on people as the dominant organism on the planet. It is also necessary to integrate the flow of information between the many institutions involved in data collection, quality control, analysis, comprehensive assessment and delivery of information to the users (managers, decision-makers, the general public), and to ensure that user needs feed back to the design of cost-effective observation and assessment programmes.

An in-depth study of Earthwatch was proposed in 1993 by the Nordic and other countries at the UNEP Governing Council to develop ways and means to strengthen the systematic observation and assessment functions of Earthwatch, and to prepare a strategy to implement the UN system-wide Earthwatch programme effectively. The same process will be required to meet UNEP's obligations as Task Manager for Earthwatch in implementing and reporting on Agenda 21. The in-depth study is allowing the agencies cooperating in Earthwatch to design a redefined and refocussed Earthwatch responding to the priorities of Agenda 21.

An inter-agency Earthwatch Working Party recently adopted a new mission statement and terms of reference for Earthwatch and agreed on a number of specific activities for its implementation.

The mission of the UN system-wide Earthwatch is to coordinate, harmonize and integrate observing, assessment and reporting activities across the UN system in order to provide environmental and appropriate socio-economic information for national and international decision-making on sustainable development and for early warning of emerging problems requiring international action. This should include timely information on the pressures on. status of and trends in key global resources, variables and processes in both natural and human systems and on the response to problems in theses areas.

Among the agreed terms of reference of the UN system-wide Earthwatch are the following:

- to facilitate access to information on on-going and planned environmental activities, and to information held by each part of the system;

- to identify possibilities for collaboration and mutual reinforcement among agency observation and assessment programmes (including GCOS, GTOS and GOOS) and reports, and with outside partners including governments, the scientific community, NGOs and the private sector;

- to promote and monitor capacity-building for data collection, assessment and reporting;

- to improve and obtain international agreement on the harmonization and quality control of data and the standardization of methodologies to ensure reliable and comparable information on the environment at the national and international levels;

- to facilitate the wider use of information and assessments from each partner beyond its own constituency in national and international decision-making processes;

- to coordinate joint reporting on broad interdisciplinary issues such as the global state of the environment and sustainable development;

- to identify priorities for international action;

- to establish joint procedures to identify the need for early warnings of emerging environmental problems and to bring such warnings to the attention of the international community;

- to share experience in applying new technologies and in increasing the impact of environmental and sustainable development information and reports;

- to assist in increasing support for observing, assessment, reporting and capacity-building activities across the whole UN system and its programmes countries.

Activities such as Earthwatch and the global observing systems respond to the needs of a number of international users of environmental such as the United Nations General Assembly, the UNEP Governing Council and the Commission on Sustainable Development, the United Nations agencies and organization, regional intergovernmental organizations, the Secretariats of international conventions, non-governmental organizations and the scientific community. For the information to be useful, it must be relevant to decision-making processes, which generally means more highly processed and assessed information with value added. Unfortunately, many decision-makers have never been trained to use scientific and technical information, so communicating the essential message is not always easy.

This one reason for the present emphasis on the development of indicators of environment and sustainable development. UNEP and UNSTAT have established a consultative expert group on the subject, and a recently launched SCOPE project is developing proposals for highly aggregated indices of sustainable development for reporting at the international level. These would measure the global impact of national activities, the sustainability of use of managed natural resources (agriculture, forestry, fisheries), the effects on nature, biodiversity and life-support systems, and the effects of the environment on humans (health, quality of life, etc.) It is hoped that such indicators will flag important trends for decision-makers. However their calculation will require sound data from programmes such as GTOS.

The initial review of Earthwatch identified particular weakness in terrestrial assessments. While there are many sectoral studies of agriculture, forests or urbanization, there has not been an integrated view of what is happening to terrestrial resources as a whole in all their complex interactions. We need enough data to be able to project trends, identify conflict and interactions, and identify limits. Ultimately, in the face of rapid population growth and high rates of consumption, we shall have to try to calculate the sustainable carrying capacity of countries, regions and even the whole planet. What are the limits that define sustainability and constrain development?

Earthwatch will focus its results more at the environment-development interface where they can be used for management. It will be forward looking to fulfill its early warning function for the international community. It will maintain an overview and coordinate the development of the full flow of information from data collection trough validation, assessment and delivery to policy-makers, with feedback loops to ensure that the process is well-focused and user-driven.

We hope to strengthen the international cases for support to the global observing systems by improving the "value added" that can come through international cooperation. For instance, forest resource information is not only of local or scientific interest, but also can provide objective information to improve the openness and efficiency of the international market for forest products. The tropical developing countries in particular are often victims of the international trade because they lack the long-term market information that would allow them to trade more wisely and sustainable. Trade mechanisms could even be designed to improve the efficiency of global forest uses and provide compensation for non-trade uses like biodiversity conservation and watershed protection. It is such linkages that can demonstrated the pertinence of observation systems to economic decision-makers.

While the resources available for Earthwatch coordination are very limited, Earthwatch can serve rather like glue to hold together and make more coherent and effective the many significant environmental observation, assessment and reporting activities across the UN system.

2.1.3 GTOS – another step toward understanding the earth system

M. F. Baumgardner

In context

As we approach the ambitious task of defining and developing a global terrestrial observing system (GTOS), it seems appropriate to remind ourselves that we are the first generation to see the Earth as a whole. In all previous generations human attempts to understand the Earth components and processes have focused on small bits and pieces of the Earth. We would cut and paste these bits and pieces together in an attempt to show how the Earth would appear if we could see it as a whole. Now the possibilities for this enterprise have been reversed. We can now see the Earth as a whole; that is, we can begin with a synoptic view and then move from the whole to the consideration of great detail of selected small portions of the Earth, but always in the context of the whole. This provides a radical departure from the approach used by all - previous generations to understand the Earth and its environment.

It also seems appropriate to remind ourselves that we and the students we teach are the first to have the tools to study the Earth as a system – to observe, image, measure, analyze and model Earth processes and interactions among components of the Earth system. Perhaps we, as noone else before us, have a better abbreviation for Chief Seattle's observation in 1854 that in the Earth environment "everything is connected!" Our challenge is to understand these connections, to measure changes which are occurring in the system and to predict both the short-term consequences and long-term implications of these changes.

GTOS in historical perspective

Many treatises have been written about the human impacts on the environment and Earth resources through the centuries. For the purpose of this paper, the focus of attention will be on the last half of the 20th Century. It is during this period that in a very real sense, for the first time in human history, environmental concerns have "reached the top of the agenda" both politically and scientifically. We are asking as never before: "What is happening to the Earth system?" GTOS is an evolving offspring of this growing concern. Let us review some of the steps along the way during recent years which have brought us to this meeting in which we are focusing our attention on the harmonization of environmental data, especially data related to terrestrial ecosystems.

Every environmentally concerned person must have a list of events which in his/her mind must be included as important in the environmental scheme of things. Any list of events which have contributed to this evolving concern for the Earth system during this half century might include any of the following activities or movements and more.

Selected illustrative list representing events, movements contributing to present concern for Global Change:

1962	Rachel Carson published Silent Sprint.
1972	First United Nations Conference on the Human Environment, Stockholm
	U.N. established United Nations Environment Programme.
	NASA launches fisst land satellite for Earth observation.
	U. S. EPA bans DDT.
1974	UNEP established the Global Environment Monitoring System (GEMS).
1977	United Nations Conference on Desertification, Nairobi.
	United Nations Conference on Water, Argentina.
1979	United Nations Conference on Science and Technology for Development.
	International Convention on Trans-Boundary Pollution.
1980	World Conservation Strategy released by environmental groups.
1983	Second United Nations Conference on Population.
	NASA established Earth System Sciences Committee.
1985	UNEP established Global Resources Information Database (GRID) programrne.
1986	World population reaches 5 billion.
	Chernobyl nuclear power plant accident in Ukraine.
	Annual meeting of ICSU launched IGBP.
	Workshop on Agricultural Environments sponsored by FAO aand CGIAR
1987	In Montreal Protocol on ozone, 24 nations pledge to halve CFCs by 1999.
	Publication of *Our Common Future* by the World Comrnission on Environment and Development.
1990	Publication of *The International Geosphere-Bioshpere Programme: A Study of Global Change*. First description of IGBP's core projects on Global Change and Terrestrial Ecosystems, and Global Change and Ecological Complexity.
1992	Earth Summit (United Nations Conference on Environment and Development) in Rio.
	Publication of *Agenda 21*
1994	United Nations Conference on Human Population, Cairo

On-going International Activities contributing to current global change thrust:

1950s– United Nations Agencies' involvement in international development and environ-
present mental monitoring: (FAO, UNESCO, UNEP, WMO).
 Creation, development and evolution of network of International Agricultural
  Research Centers (IARCS).
 Emerging support among international development funding agencies for
  environmental impact studies in all development projects.
 Increasing public awareness of environmental degradation and concern for
  human impact on the environment.
 Rapid development and activities related to the Green movement, global
  proliferation of non-governmental environmental advocacy organizations.

Events, meetings during past five years leading to GTOS:

1989– There have been numerous international meetings, studies, workshops and symposia
present which have contributed to ideas and concepts of and needs for a Global Terrestrial
 Observing System. Several of the persons in this Workshop have participated in one
 or more of these events. Only three of these UN and/or ICSU-sponsored meetings,
 studies, symposia, workshops will be listed here.

1991   International Conference on an Agenda of Science for Environment and Development into the 21st Century. Vienna.
1992   Workshop on "Detecting and Monitoring Change in Terrestrial Ecosystems." Sponsored by UNESCO-MAB, IGBP-GCTE and OSS in Fontainebleau, France, 27–31 July 1992.
Workshop of UNEP Scientific Advisory Committee for Terrestrial Ecosystems Monitoring and Assessment (SACTEMA), Prague, 9–11 September 1992.

Emerging and evolving science and technology in the last half of the 20th century which make GTOS possible:
1950–   Development of the computer.
present   Evolution of software for image processing, spatial database management.
Evolution of data storage and retrieval capabilities.
Aerospace Earth observing systems with expanding array of sensors.
Electronic and instataneous global communications.
Electronic networking and multimedia presentation and distribution of data/information .
Development and application of spatial statistics and environmental modelling.

International Earth observing actitivities

One of the relatively recent developments in Earth system studies is the joint sponsorship by multiple international agencies and organizations of studies to address issues related to the observing, measuring and characterizing ofthe major components ofthe Earth system. One of the concerns of this Workshop will be to consider the common interests and interactions among three of these jointly-sponsored activities – GCOS, GOOS and GTOS.

Global Climate Observing System (GCOS)
The Global Climate Observing System was established after the Second World Climate Conference to ensure the acquisition of the observations required to meet the needs for a) climate system monitoring, climate change detection, and response monitoring, especially in terrestrial ecosystems; b) data for application to national economic development; and c) research toward improved understanding, modelling and prediction of the climate system.

Sponsored by the World Meteorological Organization (WMO), the Intergovernmental Oceanographic Commission (IOC) of UNESCO, the United Nations Environment Programme (UNEP), and the International Council of Scientific Unions (ICSU), the program is being developed under the guidance ofthe Joint Scientific and Technical Comrnittee (JSTC), supported by the Joint Planning Office (JPO) in Geneva, Switzerland, at the Headquarters of WMO.

GCOS is currently assessing existing operational meteorological/atmosphere/climate systems in its mission to define and justify the observations which are required. Since the responsibilities of GCOS involve the entire Earth surface, the GCOS program is cooperating with the Intergovernmental Oceanographic Commission in its efforts to establish a Global Ocean Observing System (GOOS).

For the land surface, it is anticipated that the GCOS JPO will participate in the planning and development of GTOS. One of the objectives of GCOS is to address the impacts of climate change, particularly on terrestrial ecosystems and will cooperate with GTOS to obtain appropriate data.

Global Ocean Observing System (GOOS)
GOOS is being developed as a separate entity which will complement and collaborate with GCOS and GTOS. GOOS is sponsored by IOC-UNESCO, WMO and ICSU to meet the long-term needs for climate system monitoring and change detection.

Global Terrestrial Observing System (GTOS)
In 1993 a Memorandum of Understanding (MoU) was signed by FAO, ICSU, UNESCO, UNEP, and WMO to cooperate "in the scientifically based design and planning phase ofthe Global Terrestrial Observing System." This was in response to the specific request in Agenda 21 "to strengthen information, systematic observation and assessment systems for environmental. economic and social data related to land resources at the global, regional, national and local levels and for land capability and land-use and management patterns."

The MoU stated that GTOS shall have, as its long-range objective, to provide the observational framework and data basis for:

- detection and understanding of the impacts of regional and global change on terrestrial and freshwater ecosystems, including their biodiversity, as well as responses of ecosystems to such change, and of their role in causing change;

- evaluation of the impacts and consequences of global change on terrestrial ecosystems components and the environment (impacts of climate change, cycling and long-range transport of pollutants, human population dynamics in tirne and space, and other anthropogenic impacts);

- forecasting, prediction and early warning of future terrestrial changes and their impacts;

- validation of global models of ecosystem processes and change.

The MoU provided for the naming and support of an Ad Hoc Scientific and Technical Planning Group for formulating the overall concept and scope of GTOS. The 20-member GTOS Planning Group met with representatives of the co-sponsoring organizations in Geneva, Switzerland in December 1993 to organize and to formulate strategy for conducting its mission.

The terms of reference of the Planning Group are:

1. to review the work already done by previous meetings and bodies go plan a global terrestrial observing system, as a basis for completing a plan for agreement among the co-sponsors;

2. to advise the co-sponsors on the advisable scope for GTOS, based on the agreed user requirements, objectives and guiding principles, and taking into account the availability of resources and the need far cost-effectiveness:

3. to develop phased plans for the development of a scientifically sound operational observing programme for GTOS which will meet the requirements of the organizations involved and be based as far as possible on on-going monitoring activities;

4. to determine if special sub-groups or working groups are needed to prepare certain detailed elements of GTOS, and to organize such working groups to the extent that outside funding or other support can be obtained for the purpose; and

5. to prepare budget estimates for the international costs of coordinating and implementing GTOS, as well as indications ofthe requirements and costs for national participation in GTOS.

GTOS Committee on Data Management, Access and Harmonization.
One of the sub-groups, as provided for in the Terms of Reference for the GTOS Planning Group, is the Comrnittee on Data Management, Access and Harmonization (DMAH). This Committee was one ofthe special sub-groups appointed during the first rneeting ofthe Planning Group in Geneva in December 1993. The Munich Workshop in June 1994 was planned to support the missions of DMAH and UNEP's Harmonization of Environmental Measurements unit in Munich. This Workshop represents the first working meeting of the DMAH Committee. The agenda for the Workshop has been designed to address issues related to the Terms of Reference of DMAH, as follows:

1. To develop a data and information plan which considers the following issues and makes appropriate recommendations:
   a. statement of principles with respect to access and harmonization of environmental data,
   b. the required data management structure;
   c. specification and characterization of required data and information system (DIS) components such as metadata, datatypes, networking, archiving, and distribution,
   d. organizational and interfacing issues; e.g, capacity building, and implementation of reciprocity between global and national providers of datasets.

2. The plan should be fully responsive to the data requirements of the sponsors and other potential users of GTOS.

3. The plan should review existing capabilities, identify gaps and priorities for action associated with data and information management.

4. The plan should make every effort to build on existing capabilities, including those of UNEP, FAO, WMO, and other international agencies and organizations.

5. The plan should carefully coordinate with other complementary planning activities involving such organizations as GCOS, GOOS and IGBP-DIS.

2.1.4 GTOS National needs information requirements

H. Narjisse

Introduction

There is increasing concern over the emergence of potential and actual hazardous environmental issues such as global warming, desertification, biodiversity loss, air and water pollution, decline in ecosystem productivity and others. The negative impact of these environmental alterations on the health and integrity of terrestrial ecosystems, although not known with certainty, is a serious concern as it may threaten the sustainability and survival of the prevailing way of life and production systems.

Environmental problems have in general a transnational dimension and result from adverse cumulative effects of multiple and interdependent processes that are poorly understood and have impacts which are hardly predictable. The uncertainty surrounding these issues and their potentially immense socio-economic implications alerted the scientific community, public opinion and policy-makers and imposed an urgent need for a better understanding of these processes, so that their outcome can be predicted and hopefully prevented before they reach a critical or even an irreversible stage.

An in-depth understanding of these processes requires an accurate assessment of the baseline conditions of the Earth's natural resources, and an efficient system for monitoring and reporting changes in various environmental attributes and in resource condition of both managed and natural ecosystems. At present, efforts to collect baseline data on the condition of global resources, and to analyze and report on changes in those conditions at the global level, are few and impoverished. The results of these efforts remain in general unsatisfactory, as they are uncoordinated and consequently result into both unnecessary duplications and significant gaps in information.

The need for establishing a global terrestrial observing system (GTOS) is therefore clearly justified. The success of such program will depend, however, on its ability to provide the scientific community and policy makers with the needed data to assess adequately global environmental changes and at the same time to contribute to the development of appropriate and adapted technology packages, hence enhancing the problem-solving dimension of GTOS and its ability to support national planning bodies in resources management and reporting commitments under the terms of the climate, biodiversity and desertification conventions.

Designing a global terrestrial observing system is an extremely ambitious and complex task, given the multitude of terrestrial attributes to be considered (soil, vegetation, surface and ground water, fauna, air, coastal shore), the myriad of agencies and institutions involved, the wide spectrum of procedures and scales, the weak and sometimes controversial scientific foundation of the processes and principles driving their dynamic, and the high cost of monitoring activities.

Successful development of a global terrestrial observing system requires a clear definition of GTOS products users, identification of the resources and issues of concern, establishment of goals and objectives, and a formulation of a conceptual approach. This paper will focus primarily on how GTOS can meet local and regional information needs with regard to critical environmental issues. In particular, we will spell out those needs, and identify appropriate mechanisms to stimulate participation of developing nations to the monitoring network in order for them to benefit from GTOS products and contribute to the development of a meaningful database that will fulfill the needs of GTOS.

GTOS objectives and products users

GTOS co-sponsors insisted that GTOS must (i) be built up from the contributions of national institutions already in place, (ii) improve national and international decision making with regard to water and land sustainable development, and (iii) provide appropriate information to global research scientists.

From this perspective, national institutions are expected to play a pivotal role in GTOS operation. This means that GTOS should be attractive and appealing to a broad range of national governments. Unfortunately, review of existing monitoring activities around the world indicates an uneven geographical distribution coverage, with the majority of monitored sites located in North America and Europe. The concern to insure a credible and objective global representation of the entire range of terrestrial ecosystems suggests that further efforts be undertaken to include stations from the developing world, as these countries represent a diversity of biomes not covered by the existing monitoring networks. For these reasons, it is strongly recommended that developing nations participate, so that GTOS is truly global. This will require consideration of their needs in planning GTOS, and development of appropriate mechanisms to meet these needs. Although the latter are different from one country to the other, they all seek to achieve the following objectives:

1. Assessment of the condition of the nations' natural resources;
2. Evaluation of the effectiveness of current environmental policies and programs;
3. Identification of emerging environmental problems before they become widespread or irreversible and formulation of management alternatives to solve them

Improving our understanding of the global terrestrial environment, the mechanisms of its dynamic and change, and our predictive capabilities is another equally important dimension of GTOS. Many environmental problems facing mankind are indeed transnational in their nature, and require active collaboration at the regional and global level for their assessment and solution. This makes the scientific community, as represented by national and international research organizations, an important user of the data and information generated through GTOS activities.

Memoranda of understanding and cooperative agreements between GTOS and these organizations can be a possible vehicle to secure long term collaboration between these entities.

Finally, the long-term nature of GTOS and the need for continuous planning, coordination and leadership in implementing GTOS activities suggest that regional and international agencies and institutions be fully involved in the design, funding and operation of GTOS structure. The international organizations are expected to play an essential and active leadership role by providing guidance in formulating and adjusting GTOS goals, assisting in its networking activities, and securing diffusion of its products to all potential users and in appropriate format.

Information needs of developing countries

The priorities of developing nations are currently dictated by their concerns over the deteriorating standard of living of their population and the necessity of devising appropriate policies to alleviate poverty and improve the quality of social services provided to their citizens. In view of this, developing countries are more concerned with day-to-day difficulties, and they consequently allocate very few resources and little attention to planning and development of assessment tools such as monitoring.

Position of the problem

The concept of resource monitoring is new in many developing countries except for national data related to population census and various national statistics, which are, in general, regularly collected. The lack of technical staff to perform inventories and to allow ready access to remote sensing technologies, the limited financial resources and the urgency of day-to-day problems are the main constraints facing the development of environmental monitoring activities in developing nations.

This results in a lack of comprehensive inventories of natural resources, and therefore baseline information on the condition of these resources. Moreover, the data derived from existing inventories can hardly be aggregated to be meaningful, as they are fragmented, performed at different scales and with variable reliability. In addition, the few baseline information that have been collected are poorly archived, hardly accessible as they are held by numerous agencies with very few provisions for exchange and coordination between them. This leads to another problem relative to the gap existing between collecting information and assessing it, and ultimately using it effectively as a management tool.

Lack of qualified manpower to perform resource inventories is a serious issue that is unlikely to be solved in the forcible future under prevailing economic conditions of developing nations. Most inventories are indeed implemented within the framework of development projects funded through bilateral cooperation or international financial institutions. In these situations, inventories are usually components of technical assistance packages performed by foreign consulting companies or individual experts with little participation of national institutions and personnel, hence fostering the latter's dependance on technical assistance from abroad.

Tremendous efforts are therefore needed to promote the concept of monitoring as an essential tool to document the condition of natural resources and the nature and extent of land use impact on them. The adoption of such concept by developing nations requires enhancing motivation and awareness of all parties involved in the assessment, planning, management and use of natural resources. Financial assistance is also critical to initiate long-term monitoring programs and illustrate the benefits that can be potentially derived with regard to natural resource sustainable use, protection and conservation as well as international reporting commitments.

National needs and information requirements

The focus on national needs should perhaps not be on particular needs of given countries, but on national needs that are shared and considered important by several countries. The emphasis will therefore be targeted on environmental issues that are transnational and have regional relevance. Monitoring and solving such problems require necessarily the combination and coordination of regional efforts.

From this perspective, the major environmental issues reported in many developing countries include soil erosion and sedimentation, salinization, deforestation, loss of biodiversity, drought and desertification, Pollution in urban areas, land abuse and decreasing agricultural yield. In view of these issues, the most critical environmental variables to be considered relate to spatial data bases on agroclimatic conditions, landform-soil relationships, water resources and agrohydrology, land cover, land use, status and hazard of land degradation, air and water quality.

In light of GTOS objectives and the current development priorities of developing nations, assistance needs of the latter can be classified into four major categories:

1. *Assessment of the national state of the environment in relation to global processes.* This will consist of collecting, compiling, updating and distributing information on water resources, land use, land classification and condition (vegetation and soil inventories), genetic resources (particularly endangered plant and animal species), biomass and productivity.

2. *Evaluation of the on-going conservation programs in natural and managed ecosystems.* Elements of these evaluations include compiling and analyzing data on species/site interaction and ecological health, and correlating demographic, socio-economic and natural resources information at micro and macro-levels.

3. *Providing access to appropriate technologies for water and land management and sustainable development.* Governments, at the appropriate levels, should cooperate with international organizations and particularly with developed nations, with a view to enhancing transfer of technology and specialized training and ensuring access to experiences and research results obtained elsewhere.

4. *Securing financial support for monitoring activities at the national level.* External funds would be used for training, installation and maintenance of instruments and equipment necessary to perform data collection, processing, storage and exchange and to promote the flow of GTOS products to the users.

Incentives to be provided to developing countries

To make GTOS appealing to developing nations, the problem-solving dimension of this structure should be clearly stated and the mechanisms of delivering such assistance are adequately defined. Although the roots of dealing with environmental problems in developing countries are both managerial and financial, the structural facet of these problems remains the most critical constraint as it affects the human resource development component and the efficiency of environmental programs implementation. The incentives to be provided to developing countries should therefore include a set of strengthening actions aiming at up-grading national institutions's ability to perform inventory/monitoring tasks efficiently and with the required quality and to conduct and sustain viable research and development programs. The assistance needed has therefore multiple facets including policy formulation, institution building, human resource development and technology transfer.

Policy framework

Governments, at the appropriate level, with the assistance of international and regional organizations, should be able to improve their indigenous capacity to plan, develop, and implement national and regional information policy and secure long-term commitment to these activities. The systems, methodology and know how generated through the development of the baseline information necessary to undertake/update periodic inventories and surveys, planning and evaluation of on- going programs will help improve efficiency, up-grade institutional stature and facilitate projects formulation and evaluation for environmentally sound natural resources management and conservation. One way to improve the public image of agencies dealing with environmental assessment, protection and restoration is to secure appropriate training of directorate staff in managerial skills to enable them relate adequately to the local political establishment and sell their expertise and environmental programs. Long-term commitment to monitoring activities requires also that adequate efforts are undertaken to enhance public awareness and motivation, particularly with regard to environmental health and how it may be negatively impacted by land misuse. This can be achieved through promoting initiatives such as press conferences, press releases, and public symposia, with active participation of local associations, non governmental organizations and elected councils.

National institutions building

Institution building aims at creating adequate working facilities and conditions. This is necessary to ensure that data available through the network are validated and are of the required quality. Shortage of equipment, foreign currencies, literature resources, and poor dealer network are among the infrastructural constraints that need to be overcome by national institutions to attain operational status. It is unlikely that GTOS will have the financial means to meet all these needs at the national level. A significant portion of these constraints can, however, be overcome by forming an enduring, voluntary professional partnership with a peer institution in developed nations or between neighboring countries through regional network, allowing all the network members to share the infrastructural resources available at the network node. These should include necessarily the appropriate equipment and staff to operate a remote sensing and GIS laboratories, ready access to online connections and other means of communication and data distribution including electronic means. At the national level, assistance to local education institutions in establishing demonstration sites/laboratories to serve as models and training facilities is also worth being considered given the multiplier effect this may have on promoting training and research activities in these institutions.

Human resources development

Essential means for effectively promoting monitoring activities in developing countries include personnel training to allow them update their knowledge and develop appropriate skills. Usually, the most crippling infrastructural shortage in developing countries is of technically trained persons and operational technical services to secure adequate maintenance of scientific equipment. Two steps are needed: first, training must be supplied by short courses conducted in the host country or, if abroad, by persons familiar with conditions in the host country. Such training will provide guidance and expertise in field sampling efforts and measurements of various environmental attributes. In addition, it will provide hands-on experience to the trainees in the areas of planning data collection, data collection, analysis and reporting. The second essential step consists of follow-up activities that address local technical problems as they arise, with a short response time. Follow-up activities should be planned so that they do not foster dependence on the donor.

In view of the topic that we are dealing with, building human resources should devote special attention to the area of information sciences through needs-based training at all levels and, particularly, training of managers and trainers to strengthen the multiplier effect. In this respect, provision of opportunities to local scientists to keep up with advancing technology such as computerized data management, forecasting systems using telemetry, and application of remote sensing to provide, on a medium and long term scale, the necessary complements to the insufficient field measurements are topics that deserve special attention.

Means of implementation

The diversity of contexts, values, methodological approaches, environmental issues, and differential development status and priorities of nations participating in GTOS are among the constraints that will dictate a phased and flexible approach in planning GTOS. In this prospective, GTOS implementation will likely follow an extensive review of the initial strategic plan and evaluation of demonstration projects dealing with networking, training, methodology harmonization, technology transfer, and provision of financial support. These exploratory projects should be tested in various regions of the world. They can be used as a network starter in regions without operational network or as a mean to consolidate and strengthen existing networking activities in the case of regions where active networks are already operating.

Networking

Provision of the services described above can best be achieved through a regional network staffed with a coordinating secretariat acting as focal point. This unit, properly budgeted and staffed, will have the responsibility of promoting interactions and cooperation between individual data collectors through training, data distribution, meetings organization and promotion of collaborative research. It is desirable that each of these regional centers may, if needed, establish linkages with advanced national/international research/academic organizations. This will improve network access to technologies, data systems and sources relevant to supporting natural resources management, conservation and development. This will also enhance the capacity of local scientists and managers to obtain relevant information and bring about a more effective transfer of technology at the grass-roots level. In a later stage linking the regional networks into a super network will further improve sharing of data at the global level, by promoting adoption of standards, compatible and harmonized methodologies, and transfer of technologies adapted to developing nations' environment.

Organization and operation

Each participating country will have a coordinating unit where all environmentally involved disciplines and agencies are represented. These units will be used as a support for interagency communication, synergism and exchange and will serve as a counterpart to the regional network. The host country unit should be equipped, in addition to computing facilities, with telecommunication systems allowing regular courses broadcast, and interaction with the regional node with an affordable cost. Availability of such systems has another advantage, as they will allow involvement of as many local experts as reasonable to facilitate buy-in of the programs by local and national institutions. Over time, operational costs of such units can be covered through a system of pay-for-use basis.

With regard to data collection and mapping, these can best be done by nationals, coordinated and assisted by the regional center that will have the responsibility to ensure that data are collected with the required quality and made available through the network. Digitalization and interpretation cost may be reduced by centralizing and consolidating interpretation centers where equipment and skills are shared.

Financial support

The financial implications of achieving GTOS goals are immense and probably beyond the scope of its tight budget. On the long-term, sustaining GTOS activities will depend on national governments commitment to monitoring and environmental research and development activities. This means that at some point, success in the adoption and implementation of GTOS programs should be linked to the general economic conditions of member nations. In this situation, promoting and strengthening of national monitoring capabilities are areas that should be dealt with in conjunction with the more general socio-economic development framework. For these reasons, we suggest that monitoring activities become component of environmental projects to be funded through phase II of GEF.

Vision of future GTOS

As stated earlier, in view of GTOS complexity, its implementation should necessarily be flexible and performed gradually on the basis of predefined phases. GTOS may therefore be planned in two successive phases lasting, for example, 5, and 30-year periods. Each of these phases will pursue clearly defined goals that may evolve through time and that should consequently be reviewed and adjusted as necessary.

The 5-year goal may be to establish baseline conditions, develop GTOS management structure, secure long term national and international financial commitments, and assess the ability of GTOS to integrate information to determine regional ecological condition. Throughout the first phase, we expect regional networks to be put in place, and to achieve significant progress in capacity building of developing countries. In addition, mechanisms for improved electronic communications between and within networks as well an adequate training of directorate staff, managers and technicians be identified and necessary measures be undertaken to implement them. Similarly, at the end of the first phase of GTOS implementation, public awareness and indigenous capacity of developing nations to deal with environmental problems, including monitoring, should attain a satisfactory level. Throughout this period, efforts will be undertaken to promote initiation of joint ecological research to assess regional trends in the condition of selected ecosystem resources and attributes, and develop test scenarios to determine causes of regional alteration, degradation, or enhancement. It is expected that the links to be established between the scientific community and the regional networks will further up-grade human resources development in developing countries and lead to the formulation of a unified terminology and a harmonized methodology.

The 30-year goal aims at consolidating the achievements of the first phase. Particularly, the various links formed throughout this phase will contribute to a better understanding of ecological processes and functions, their impact on ecosystem health and integrity, and the production of validated technology packages susceptible of providing accessible and appropriate managerial and technical solutions to stop environmental degradation. The collaboration between the regional centers and the scientific community can also lead to the development and testing of new monitoring approaches in light of the technological progress in the form of global positioning system, remote sensing, and geographic information systems and along with the new emerging disciplines of landscape ecology, restoration ecology and economic ecology. This phase aims at producing the appropriate models to predict global change and the adequate technologies for the restoration of degraded terrestrial ecosystems and for their conversion to ecologically adequate desired condition. The training and support for human resources development and public awareness provided through phases I and phase II will enable agencies and public opinions of developing countries to attain the appropriate maturity status in dealing with environmental issues. This is necessary to sustain national monitoring programs and associated activities.

Conclusion

Monitoring activities are presently restricted primarily to developed nations in north America and Europe. In contrast, very few sites are regularly monitored in developing nations, despite the fact that the latter occur in a variety of terrestrial ecosystems and comprise a significant proportion of the surface area of the globe. Given the transnational nature of most environmental issues and concerns, the inclusion of sites from developing countries is essential to the understanding of environmental disturbances, and to the elaboration of appropriate models to predict global changes.

Definition of a conceptual GTOS approach is an immense and complex task, as this structure will confront extremely diverse situations and challenges in both their ecological and socio-economic characteristics. The difficulties associated to the design of GTOS structure can be overcome by adopting a flexible and phased approach, promoting a spirit of collaboration between participating institutions and individuals and gaining long-term commitments of the donors.

Incentives should be provided to developing countries to insure their participation to global assessment of the environment. Among these incentives, assistance in building of human and institutional capitals and increasing public awareness is critical to raise the level of environmental literacy and secure long-term commitment of developing nations to monitoring activities. Provision of such incentives will facilitate technology transfer and assimilation and enhance political support necessary to sustain environmental monitoring activities.

Achieving GTOS ambitious goals will also depend on the long-term availability of funding, the firm commitment of participating countries, the extent of program integration and the dedication of GTOS managing structures at the national, regional and global levels. The sustainability of monitoring efforts and environmental protection and restoration programs in developing nations cannot be expected, however, unless the latter achieve significant progress in the area of social and economic development. For this reason it is essential that GTOS activities are closely coordinated, throughout the two phases of their implementation, with donors and lenders at the bilateral and multilateral levels.

2.1.5 Harmonization of environmental data: the requirements for developing a consistent view of the environment world-wide

H. Keune and P. Mandry

Introduction

Environmental problems are increasingly seen to derive from, or have an impact on, areas far beyond the strictly local. Many problems are now recognized to be essentially global in nature, i.e. either occurring on a global scale (e.g. climate change), or sufficiently widespread to have planetary influence or significance (e.g. gaseous emissions, water pollution, air pollution, waste disposal). Increasingly decisions are having to be made locally which require information from a large number of disparate sources, across a multitude of disciplines and covering a broad geographic scale. Similarly sustainable development of the global environment requires environmental policies, which necessitate having reliable information on the state of the environment and trends. The multinational and interdisciplinary nature of many problems means that the data used in environmental assessments will come from many different sources. Data from different sources, however, can only be integrated and aggregated meaningfully if they are comparable and compatible and of known and defined quality. Data are basic elements of information. Understanding information leads to what we call "knowledge" (Fig. 2.1.5.1).

There is a large amount of environmental data available. But often the data are of unknown quality, are not accessible, or those who need it do not know of their existence. Furthermore

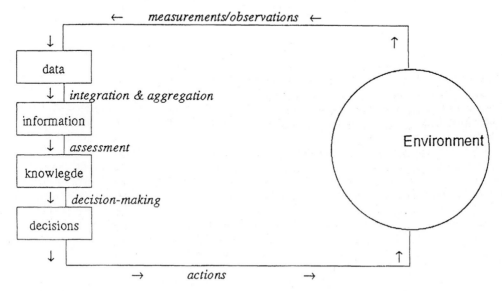

Fig. 2.1.5.1:
General Scheme for the Development of Environmental Information for Decision-Making

much relevant data are sparse, incomplete, unbalanced or non-representative. And those which are available are often in a form which does not allow meaningful aggregation, i.e. are not comparable and compatible.

The present situation has arisen partly because there is no consensus on what should be observed, measured and/or monitored and how, or in what form it should be recorded or archived. It is the task of global harmonization, and thus for the UNEP-HEM office, to look at ways of overcoming this situation. Harmonization is a means to an end, not a goal in itself. It is the 'tool kit' for ensuring that data from different sources can be meaningfully combined. Harmonization is a concept which can be applied at many different levels and in many different ways, from agreements in data transfer in a local area network to agreements on a standardized global reporting system for meteorological observations. A genuinely global view of the state of the environment, however, can only be developed on the basis of data which are harmonized, and thus compatible, at a global scale. Harmonization within specific programmes at the global scale is often not sufficient, it is also necessary to ensure that the data from these programmes are compatible and comparable with each other. The same variables are often investigated in widely differing programmes, equally different programmes may gather information relevant to quite different sectors. Harmonization must ensure not only that data collected in different programmes are comparable and compatible, it should also ensure that integration and aggregation of data are performed at different places in such a way that the results are still comparable at the level of environmental information and that assessment is done in a comparable fashion so that a common understanding about the state of the environment can be developed at all levels. It will only be possible to develop the common information needed as a basis for making decisions on sustainable development if data collection, integration, aggregation and assessment are harmonized.

Essentially two approaches can be recognized towards developing a consistent picture of the environment on a global scale, i.e. obtaining (sufficient) comparable and compatible data across the globe. One method is to develop effective global networks to monitor and detect current status, trends and change in environmental systems using representative sites or areas. The other is to develop recommendations and work towards global agreement on the choice, recording, archiving and dissemination of the variables needed in environmental assessment. These approaches are complementary rather than exclusive. Essentially, they embody two slightly different approaches to harmonization, and by considering these approaches in turn the requirements which must be fulfilled to enable harmonization can be recognized.

The first approach is the development of global monitoring or observing networks. Of the three global environmental observing systems which have been, or are in the process of being, established (the Global Climate Observing System [GCOS], the Global Ocean Observing System [GOOS] and the Global Terrestrial Observing System [GTOS]), GTOS is the one most likely to focus on representative sites using agreed protocols for observation and recording as the main tool for obtaining consistent data sets. Even so, GTOS is intended to build on existing activities rather than establishing a new monitoring programme, which considerably complicates the harmonization task. For GTOS to be successful, it will be necessary to develop strategies for the timely and efficient flow of credible, useful environmental data along the path from the time of data acquisition to information delivery. In order to know what relevant data, or sources of data, already exist it will be important to know who is observing or monitoring what, where, when, how and why. That is, it would be useful to have an inventory of measurement techniques, terminology, classification systems and programmes within GTOS, links with GCOS and GOOS, and other relevant data sources and related monitoring and research activities. Cross-sectoral or integrated monitoring will be required to obtain a complete picture of the state of an ecosystem, and a plan will be needed showing the requirements for such an integrated approach and the links between different data sets. Equally a means must be found for ensuring the comparability and compatibility of data obtained in the network. In other words, GTOS needs a data and information plan (UNEP, 1994), based explicitly or implicitly on a framework showing the links between data sets, how and what data will be collected, integrated, aggregated and assessed, the needs for harmonization, potential sources of data, and pathways for data flow. Once it is established, GTOS is expected to provide harmonized data sets covering an agreed set of variables from a selected range of sites across the globe.

The second approach has broader implications since it is concerned with improving the comparability and compatibility of environmental data world-wide, and not just from a selection of representative sites. For harmonization on a global scale, it is necessary to develop a means of cross-linking and co-ordinating individual efforts aimed at harmonization within specific programmes or scientific fields. Harmonization strategies must be implemented which ensure not only that data collected at any location in the world is comparable and compatible with data on other variables, but also to ensure that the integration and aggregation of data is done by comparable methods and that the necessary assessments use accepted and compatible models. This requires the development of a conceptual framework for environmental data and information systems that can accommodate the results and measurements from all monitoring and research programmes in a compatible manner. This framework should make it possible to recognise cross-sectoral linkages, the overlap in interest areas of different programmes and activities, and the interconnection between activities at different levels and in different fields. The framework needs to cover all aspects necessary for the production of environmental information from data collection to total assessment. It would thus provide the basis for all specific harmonization and standardization activities, and a means of identifying possible areas of conflict and of ensuring co-ordination of approaches. Inasmuch as the world can be considered to be a single giant ecosystem, such a framework would effectively be a summary of the cross-linkages within this system.

Thus it is clear that these two different approaches actually have rather similar requirements. Both need information on existing activities and sources of data, both need a framework showing the linkages between these activities and data sets. A framework showing the data requirements for assessment of terrestrial ecosystems and the existence of and linkages between sources of data as proposed for GTOS, could be seen as the core of the framework needed for broader global harmonization efforts. There are considerable advantages to be gained from focusing efforts in the first instance on terrestrial ecosystems: the prospective task becomes more manageable; approaches to harmonization can be developed in a limited field and tested in practice; the data sets required for assessment of terrestrial ecosystems cover a wide range of sectors and will be relevant in many other fields.

Definition of Harmonization

Harmonization in the context of environmental monitoring seeks to bring together various kinds, levels, and sources of data in such a way that they can be made comparable and compatible and thus useful in decision making. Harmonization differs from standardization in that it does not impose a single methodology, but rather seeks to find ways of integrating or making „an agreeable effect" from disparate methodologies. The actual harmonization task differs according to the level of data collection, data integration and aggregation.

Harmonization is involved throughout the process from observations to policy setting, i.e. from data collection, to development of information by data integration and aggregation, and formation of environmental knowledge as the common basis for decision-making. Effective harmonization requires good knowledge of the nature of the phenomenon being measured as well as a clear concept for what is to be achieved.

Depending on the level in the hierarchy the harmonization requirements are:

a) **Data collection**
   Agreements on
   - catalogues of variables and choice of parameters,
   - site selection, measurement protocols, survey design
   - QA/QC procedures for sampling and analysis and "codata" requirements.
   - measurement techniques and technology

b) **Integration and aggregation of environmental data**
   Agreements on
   - globally accepted terminology,
   - scales and working levels,
   - classification systems,
   - meaningful indicators and reliable models,
   - GIS procedures, spatial frameworks, and spatial data aggregation.

c) **Assessment of environmental information for decision making**
   Agreements on
   - statistical methods,
   - scenario techniques and expert systems,
   - simulation models and evaluation procedures
   - sustainable development models and cost benefit principles and models.

These basic principles apply whether considering a limited network of sites, a global network of sites or broader global harmonization.

Harmonization within the Global Terrestrial Observing System (GTOS)

Monitoring terrestrial ecosystems is different from the more continuous systems of the atmosphere and oceans. The components of terrestrial ecosystems are heterogeneous, and internal linkages are complex in themselves, as are interactions with the atmosphere, hydrosphere and lithosphere. Terrestrial observables such as soil, vegetation, flora, and fauna vary rapidly with both gradational and abrupt discontinuous boundaries. In a statistical sense there is much more limited spatial auto-covariance, i.e. an observation at one point is often a poor predictor of the condition at even nearby points. This leads to more complex questions of what is appropriate to measure, and how the results of localised observation can be spatially aggregated and summarized.

As mentioned above, in order to develop a concrete data and information plan for a global terrestrial observing system it is necessary to have, implicitly or explicitly, a conceptual framework for terrestrial monitoring on a global scale. This conceptual framework is the basis for recognising the linkages, overlaps and requirements of the data needed in order that the network objectives can be fulfilled. A useful starting point for considering the development of such a conceptual framework can be found by looking at the progress made in developing cross-sectoral approaches at a local or regional scale in a number of programmes concerned with ecosystem monitoring and research.

The approach to environmental monitoring and research has changed considerably in recent years. The emphasis on sectoral assessments is giving way to a more integrative approach. Studies of forest damage and chemical pollution, for example, have shown that changes in the environment cannot be understood in a medial, sectoral or compartmental context. Only an interdisciplinary approach can lead to a better understanding of the complex processes that take place in the environment. Scientists in many countries are convinced that environmental monitoring should not be concentrated on single environmental compartments (atmosphere, vegetation, soils etc.); similarly an integrated approach seems to be much more suited to environmental policies which aim at global sustainable development. The move towards an integrated approach is equivalent to a move towards ecosystem assessment, rather than separate assessment of air, soil, plant etc. compartments.

The fundamental concept underpinning integrated (interdisciplinary) approaches is the ecosystem as a describable entity. 'Ecosystem' is understood to mean natural and/or anthropogenous landscape units which contain abiotic and biotic structures functionally correlated and governed by site characteristics and (long term) management practices (e.g. agricultural and forest ecosystems, woodland ecosystems, water landscapes, urban ecosystems). There is a hierarchy of levels of resolution above and below any particular ecosystem. Ecosystems and components of ecosystems interact in both temporal and spatial dimensions. In general, integrated ecological monitoring is a combination of spatial monitoring of (terrestrial) ecosystems and ecosystem research. It studies the conditions under which terrestrial and aquatic ecosystems are stable, what impacts are to be expected as a result of their exploitation and anthropogenic environmental changes, how their cultivation can be made ecologically sound, which measures need to be taken to ensure their redevelopment and lasting protection, and what social and economic consequences these will have. The basic approach is to establish a number of representative sites for integrated monitoring and research, and then extrapolated the results obtained at these sites to larger areas or to distant but similar ecosystems.

In order for this approach to be successful it was necessary to:

– establish a methodological framework for integrated analysis of the data collected,
– define a core set of monitoring data,
– select and establish representative sites for monitoring and research;
– co-ordinate and standardize (harmonize) scientific and monitoring efforts at the different sites.

These are then some of the steps which need to be considered when developing a global framework for an integrated approach.

The integrated approaches to environmental assessment which have been developed in a number of programmes at a local scale in the fields of ecosystem monitoring and research, will provide a useful starting point for considering the development of the integrated global framework implicit in the development of the GTOS data and information plan.

References

Environment Programme for Europe, "Environmental information" a Discussion Paper prepared by the CEC, DG XI EEA-TF, 1992.
Long-term Ecological Research. An international perspective. SCOPE 47. Ed. Risser P.G.. John Wiley & Sons, New York, 1992, pp 293.
Conceptual Framework for Research on Global Change. The Federal Minister for Research and Technology. Bonn. 1992. pp. 77.
UNEP (1994) Report of the First Meeting of the Ad Hoc Scientific and Technical Planning Group for a Global Terrestrial Observing System (GTOS) 13–17 December, 1993, Geneva, Switzerland. Nairobi: United Nations Environment Programme. Ref. UNEP/GEMS Report Series No. 24.

2.2 Disparities in Sampling, Parameters and Metadata: Environmental Monitoring and Assessment as a Unifying Basis

W. Schröder and O. Fränzle

The Problem

The sustainable development of man's environment requires environmental policies. A prerequisite for decision making is the assessment of observations. Observing produces data which have to be of defined quality to know the extent to which they represent the "true" reflections of the observed reality. Data are basic elements of information. Understanding information is what we call "knowledge". Today there is a great amount of environmental data available. But often these data are of unknown quality, they are not accessible, or those who need them do not know of their existence. Furthermore many relevant data are missing. This is especially true for terrestrial ecosystems on the global scale.

Among other things this unsatisfactory situation is due to the missing consensus on what should be monitored and how. In short: There are many disparities in sampling, parameters and metadata. To improve the situation it is therefore necessary to develop a conceptual framework for environmental monitoring and assessment as a basis for harmonization needs. It should be based on ecosystem research and consist of procedures for (cf. I):

a) assessment of environmental data
b) data collection by means of environmental monitoring and research
c) integration and aggregation of environmental data by means of:
   (i) statistical methods, and
   (ii) simulation models, and/or scenario techniques.

These requirements are fulfilled in the *Ecosystem Research in the Bornhövel Lakes District, Germany* (cf. II).

I) Requirements for Environmental Monitoring and Assessment of Terrestrial Ecosystems

Assessing Ecosystems

Man, animals and plants are elements of the biosphere. Together with the atmosphere, hydrosphere and litho-/pedosphere the biosphere builds up a system that we call "the world" or "man's environment". The spheres are linked together at any scale (local, regional and global) by fluxes of energy and matter, and the whole of both these systems and their interactions is called the "oikos". In terms of system theory the oikos is an ecosystem which can be subdivided at any level of analytical hierarchy into subsystems. Ecosystems consist of biotic and abiotic elements. Biotic elements are: plants, animals, and man. Abiotic elements are: water, air and soils. Between all these elements there are many relationships; they determine structure and function of ecosystems.

Sustainability of the world's ecosystems is the target of environmental policy. Sustainability requires precaution which implies that adverse impacts resulting from human activities are predictable. Thus prediction of adverse impacts on ecosystems must be based on the current knowledge of ecosystems. The expected reactions of ecosystems have to be the strict standard for:

- the assessment of ecosystems with respect to observed impacts and defined criteria for environmental quality, and
- actions as part of environmental policy, environmental law and environmental protection.

The predicted possibility of adverse impacts is considered the basis for making provisions (GENTILE & SLIMAK 1992; MESSER 1992).

German Environmental Monitoring: Paradigm of a Framework for Global Harmonization of Environmental Monitoring?

Because ecosystems are linked together at all scales, global environmental assessment and management should be based on a framework of ecosystem monitoring and research. There are many concepts of environmental monitoring. In Germany most scientists are convinced that environmental monitoring should not concentrate on single environmental compartments (atmosphere or vegetation or soils or ...); on the contrary an integrated, i.e. ecological approach is much more appropriate in terms of environmental policy which aims at global sustainable development. Therefore in Germany environmental monitoring and assessment integrates:

- *Ecosystem research* (A): spatially oriented geo- and bioscientific research to understand the function and structure of ecosystems;
- *Environmental monitoring* (B): spatial monitoring of representative types of ecosystems and their compartments, to corroborate an ecosystem-oriented environmental assessment and management, and
- *Environmental specimen banking* (C): storage of representative environmental specimens for documentation and retrospective analysis.

Thus the German environmental monitoring and assessment approach is a combination of comparative bio- and geoscientific research and monitoring of terrestrial ecosystems. Ecosystem research and monitoring is the ultimate basis for environmental assessment. The assessment of environmental quality then is a prerequisite for environmental policies, environmental law and applied nature protection. A sustainable protection of ecological structures and functions has to follow defined criteria for the assessment of the actual and possible risks which involves:
- the analysis of the situation by means of (A), (B), and (C) and
- its subsequent comparative evaluation.

Collecting Data on Ecosystems Structure and Functions

Environmental monitoring and assessment starts with defining
- the objects of interest,
- the methods of data collection and quality assurance, and
- the methods of data evaluation by means of statistics, simulation models and scenario technics.

Environmental data are collected by measurements within the framework of (A), (B), and (C). Measuring consists of observing objects and their relations and representing these structures on analogous relationships between values (categories) of variables. The validity of such a projection depends on the following:

a) representativeness of the observed objects/elements which constitute the random sample (sound reproduction of the population by the random sample),
b) selection of object attributes relevant for hypothesis testing and selection of specific methods for detection of these attributes (meaningfulness and comparability of the test/survey) and
c) the main criteria of test/measurement theory: objectivity, reliability, validity (precision of relation between a random variable and its measurement signal) (FRÄNZLE 1984, 1987, 1990, 1994; REAGAN & FORDHAM 1992; SCHRÖDER et al. 1992; VETTER et al. 1991; WARD 1992).

If these criteria are not fulfilled at the lowest spatio-temporal level of environmental monitoring, this will certainly affect any other level of data aggregation. Therefore harmonization efforts have to concentrate on these criterias, and quality assurance and quality control (QA/QC) mechanisms are essential. QA/QC procedures have to be integrated as standing elements in every environmental monitoring activity. They have to be recognized from the very beginning with the selection of objects for investigation up to the documentation of data:

i) *preparatory quality control:* selection of representative objects and decision for methods of examination,
ii) *quality control of measurements:* routine laboratory quality control and inter-laboratory quality control and
iii) *quality control of data:* plausibility check and evaluation of spatio-temporal validity of the measurement results (GÜNZLER 1994; SCHRÖDER et al. 1991).

Data Handling by Means of Statistics and Simulation Models

Evaluation of (primary) environmental data requires a set of co-data (secondary data). Co-data are: characteristics of objects under investigation (sampling sites, ...), description of sampling procedures, levels of (dis)integration of measurement with respect to space and time, QA/QC-data. These co-data have to ensure the compatibility and comparability of the primary data. In the next step data have to be integrated and aggregated. To this end it is useful to build up a database in combination with a geographical information system (GIS). Integration of data is to concentrate data about several variables measured on one object. This can be done by means of multivariate statistics and GIS-procedures. Data aggregation means concentrating data in terms of space and time by means of descriptive and multivariate classification and time series-analysis (DOWNING 1991; DUARTE 1991; MEENTEMEYER & BOX 1987).

Ecosystems are open systems with complex structures and functions. Because of complexity their reactions to stress are difficult to predict. But any assessment of ecosystems must be based on predicted stress reactions. So we have to build at first conceptual analogue models about interactions in/between ecosystems which are virtual reproductions of reality. They are the framework within which the collected, evaluated and integrated/aggregated data are then combined by mathematical functions. Digital simulation models are very helpful tools for predicting the further development of ecosystems under defined conditions (ROBERTSON et al. 1991). In terms of sustainability predicted reactions of ecosystems are normative for environmental policies and management.

II) Harmonization of Environmental Monitoring within a Global Network: Steps for Achieving this Goal

The world's terrestrial ecosystems are subjected to environmental strains of an unprecedended scale, both in rate and geographical extent. Therefore there is an urgent need for a global field-based network of ecosystem research sites. More than eighty international organizations and programmes are involved in global environmental monitoring, data managing and harmonization. So the essential elements for a global terrestrial observing system exist, but no existing network has a fully established, integrated and comprehensive programme of measurement and data management. Although there are similarities between networks, no generally agreed common set of variables and protocols exists. Although there are many candidate sites, the selection of sites within existing networks has to be based on scientific rationale and pragmatism (LINTHURST et al. 1992; MILNE 1992).

MAB Biosphere Reserves as a Potential Basis for the Harmonization of Monitoring of Terrestrial Ecosystems

The only network which is truly global in extent is the network of biosphere reserves, organized through the *Man and the Biosphere* (MAB) Programme of UNESCO. Today there are about 310 biosphere reserves. They are distributed throughout the biomes of the world although not constituting a spatially valid sampling regime. Therefore any terrestrial research and monitoring programme will have to touch upon and complement relevant climate programmes. So the task is to select research sites which are representative for the world's terrestrial ecosystems and which satisfy relevant infrastructural criteria. In Germany the ecosystem research programme was established according to these principles. One of the relevant representative research areas is the Bornhöved Lakes District.

The *Bornhöved Lake District* as a Possible Site in a Global Ecosystem Monitoring Network

After a ten years' preparatory phase the project *Ecosystem Research in the Bornhöved Lake District* was inaugurated on April 18, 1988. Since then more than 100 scientists and 20 technicians cooperate in the project. The interdisciplinary investigations of the Universities of Kiel and Hamburg, the Max-Planck-Institute for Limnology in Plön, the German Meteorological Survey, the Fraunhofer Gesellschaft, and the Biologische Bundesanstalt Braunschweig, are supported by the Federal Ministry of Science, Education, Research and Technology and the Land Schleswig-Holstein. According to present plans a minimum project duration of 15 years is forseen as depicted in the timetable of implementation. As soon as possible the present Project Center is to become the nucleus of a major Ecological Research Center at the Kiel University.

Selection of the Research Area

In the framework of the German MAB program a comprehensive long-term research scheme has been developed which aims at the comparative analysis of both structures and functions of ecosystems. To this end comprehensive areal data derived from digitalization of a series of ecology-related maps have been analyzed by means of multivariate statistics. The selection followed an operational definition of representativity. In compliance with the suggestions of an expert group this led to the definition of prioritarian study areas for ecosystem research purposes, two of which are the Bornhöved Lake District and the Berchtesgaden Alps.

Project Objectives of the Bornhöved Research Scheme

The main research objectives are the definition and modelling of structure, dynamics, and stability or resilience conditions of the closely interrelated terrestrial and aquatic ecosystems of the study area 30 km S of Kiel (Schleswig-Holstein). Focus is on ecosystem functioning under different intensities of human impact and the development of socio-economic models. Owing to the very high structural diversity of the largely anthropogenic landscape with a high number of terrestrial and land/inland water interfaces ecotone research plays a particular role. It aims at defining and modelling the structural, energetic, biotic and biogeochemical character of the various types of ecotones.

Advantage is taken of the manifold project opportunities for the institutional development of ecosystem research in Germany involving interdisciplinary training of students on the undergraduate, graduate and postgraduate levels. Formal and non-formal on-site training for young scientists, particularly from developing countries, can be arranged for. Mechanisms for a widespread valorization and dissemination of the results obtained are provided for the general public and both local and regional planners, for administrators and farmers including agroforestry management advice.

Degree of Interdisciplinarity and Plan of Research

A hierarchically structured system of several hundred working hypotheses on five levels of generality forms the basis of interdisciplinary cooperation. It groups scientists of following disciplines: agricultural management, chemistry, climatology, computer science, geobotany, geomorphology, human geography, hydrogeology, ichthyology, limnology, mathematics, meteorology, physical geography, planktology, plant nutrition, plant physiology, soil science, and zoology.

The general aims and ends of the long-term ecological research project are:
- defining and modelling structure, dynamics and stability conditions of the interrelated terrestrial and aquatic ecosystems of the study area in terms of site characteristics and biocoenotic diversity, natural and anthropogenic fluxes of energy and matter, productivity and land use;
- determining and modelling environmental strains and resilience mechanisms of the ecosystem compartments affected by disturbances with particular reference to chemical inputs.

In view of the specific structure of the study area particular conceptual attention is focussed on:
- modelling of biotic, energetic and material exchange processes between neighbouring ecosystems of different land use and fishery patterns;
- modelling site-dependent relationships between lakes and their drainage basins with focal reference to the role of different land/water interfaces;
- modelling of agro- and pasture ecosystems in the light of national and international production and marketing regulations;

- inquiries into the efficiency of environmental protection and conservation measures;
- testing the validity of spatial extrapolations of simulation models by means of comparative site analyses, geostatistics and geographic information systems;
- the paleoecological and historical reconstruction of ecosystems evolution in the study area since the end of the Weichselian Glaciation.

Areas of Application / Contacts with other Ecological Research Institutions

At *the national level* close scientific contacts exist with the Ecosystem Research Centers at Bayreuth, Göttingen, Munich/Neuherberg and Leipzig-Halle which form the *Terrestrial Ecosystem Research Network* (TERN). Further cooperation exists with the MAB 2, 6 and 11 projects at Berchtesgaden, Berlin, Frankfurt, Osnabrück and Saarbrücken, and with the National Parks Nordfriesisches and Niedersächsisches Wattenmeer (Tönning, Wilhelmshaven). The project contributes to the German Environmental Specimen Bank (Jülich) and the Ecological Monitoring Project (Kiel). At the *international level* contacts exist and are intensified with centers of ecological research in China, the Czech Republic, Denmark, France, the Netherlands, Russia, U.K., and U.S.A..

Ecosystem research in the Bornhöved Lake District contributes to the following MAB project areas:

(2) ecological effects of different land use and management practices on temperate forests;
(3) impact of human activities on grazing lands;
(5) ecological effects of human activities on the value and resources of lakes and rivers;
(9) ecological assessment of pest management and fertilizer use on terrestrial and aquatic ecosystems;
(13) perception of environmental quality;
(14) research on environmental pollution and its effects on the biosphere.

Links with the *International Geosphere-Biosphere-Programme* of ISCU have also been established. Contributions relate to the following basic themes judged crucial in the development of IGBP:

(1) change in atmospheric chemistry associated with modifications of terrestrial ecosystems and biota,
(2) ecosystem processes, and
(3) survey, monitoring and inventory of terrestrial ecosystems.

Diffusion of Scientific and Technical Information

Mechanisms for the direct valorization and dissemination of the results of ecosystem research are provided for:

- the Governments of the Federal Länder Schleswig-Holstein (cf. a, b) and Brandenburg (cf. c),
- the Federal Government of Germany (cf. d, e, f),
- the European Union (cf. g), and international organizations like UNEP or FAO (cf. h).

a) Elaboration and test of a concept of an ecologically oriented monitoring system for planning purposes, exemplified by case studies in Schleswig-Holstein

The concept of an environmental monitoring system is based on a matrix of potential land-use conflicts. They result in a site-specific manner from competitive or contrasting land-use orientations in the different fields of agriculture, forestry, water management, tourism, waste disposal etc. Following the precise definition of such conflicts, the rational approach to solution strategies involves a regionalized analysis of potential impact effects in terms of site sensitivity, existing land-use practices, and land-use patterns envisaged.

To this end data banks summarized in a meta-data bank and monitoring networks were established, assessment and simulation models developed and validated, and a geographical information system adapted in compliance with the specific requirements of different planning levels. A critical assessment of the qualities of successive ARC/INFO versions is given, followed by reflections on juridical and technical aspects of the implementation of the monitoring and planning systems suggested (FRÄNZLE et al. 1992).

b) Regionally Representative Soil Sampling as a Part of Environmental Impact Analyses

An environmental impact statement is a summary of all the information gathered on each potential environmental impact that might result from a given development proposal. It is an integral part of the procedure of environmental impact analysis/assessment. Within the framework of environmental impact analyses and assessments for numerous power-stations, waste incinerators and other industrial plants in Germany, regionally representative soil sampling has been performed by the Department of Geography at Kiel University. An evaluation of digitized maps on soil and vegetation cover as well as land-use patterns leads to a sensitivity analysis of the surrounding area. Additional inputs of pollutants by the respective plants are calculated on the basis of climatic and emission data and displayed in map form. The superimposed information layers are analyzed by means of frequency statistics, cross tabulation and neighbourhood algorithms. These procedures yield the number and location of most appropriate sampling sites for soil chemical analysis. Meanwhile, this statistical approach of regionally representative sampling is considered standard procedure for soil investigations by the Schleswig-Holstein Ministry of Nature and Environment.

c) Soil data for models and resource management at different scales

Soil data are necessary for several scientific and practical purposes in land use and environmental protection but they are hardly available in form of uniform basic data. So there is an urgent need to develop generalization methods that offer a possibility to make reliable statements about soils and soil associations (mapping units) from limited data sets. A study to reach this aim was performed at three different scales: (a) test site 1:2 000 (176 sampling points with 4 repetitions), (b) Federal State of Brandenburg 1:200 000 (60 196 grid points of the digitized soil map), (c) Germany, New States 1:750 000 (12 300 grid points).

The differential analysis of the test sites gives a ranking of parameters in terms of variability which increases from the stable parameters like texture and relief, semistable parameters like organic matter content and pH, to variable parameters like nutrients. Soil parameters have a certain structure which allows to develop parameter models (for stable and semi-stable parameters) and dynamic pedotransfer functions which permit methods to determine soil data from basic data. Parameter models and pedotransfer functions allow to generate estimates and to extrapolate them as regionalized data to areas. This complex procedure yields data sets which are much are more precise than mean values or data derived from single profiles.

The objective of the interpretation of the digitized soil maps was to select representative soils, i.e. soils defined by means of frequency statistics, in terms of pedochemical relevance, and structure of soil associations. Soil data of associations are regionalized soil parameters based on the catena principle. Thus the combination of values compiled statistically from all regional data and the derivation of subsidiary summary data from parameter models yield reliable estimates for the above scales (SCHMIDT et al. 1995).

d) Regionally Representative Selection of Soil Samples for the German Environmental Specimen Bank

Digital evaluation of geoscientific maps in various scales and subsequent multivariate data analysis form the basis of the comparative selection of soil specimens for the environmental specimen bank of the Federal Republic of Germany. The optimum location of sampling points was finally determined on a 1:1 000 000 soil map as well as on larger-scale maps. In addition an index of representativity was developed which relates to frequency distribution of the soil taxa at 2 485 reticulate grid points and subsidiary spatial characteristics (VETTER 1989).

e) The Aral Sea-Project groups about one hundred scientists from Kazakhstan, Russia, Uzbekistan, and Germany for assessing the ecological situation in the Aral Sea depression which has undergone disastrous changes under human impact during the last 40 years. Research topics are:

– paleolimnology and -ecology,
– modelling of present-day ecosystems,
– problems of prognostic modelling,
– scaling problems in GIS development and application, and
– landscape evaluation and hazard assessment.

f) Joint German-Russian project on soil analysis, soil quality assessment and soil protection

On the basis of a common assessment strategy two sampling areas in Russia (Ostashkov, Tcheljabinsk) and Germany (Bornhöved, Stollberg) were selected for comparative analyses of soil quality. The data form the basis for the development of a generalized sampling and assessment approach.

g) Selection of reference soils for chemicals testing in the European Union

Based on multivariate statistics of binary and metric data relating to the soil cover of the European Union five regionally representative reference soils (EURO-soils) were identified for chemicals testing in the EC. The soil material sampled was treated and prepared according to OECD Test Guideline 106 and analyzed in detail. The homogenized specimens were subject to a EC-wide ring test to evaluate the feasibility of the modified guideline and to validate the physical-chemical amenability of the reference soils for sorption tests. The results proved the validity of the soils selected for assessing the potential behaviour of new chemicals in soil on the basis of a comparative evaluation of the individual test results obtained.

In the light of this parametric assessment potential test soils were subsequently identified in the individual EU member States which correspond as far as possible to the above reference soils in terms of both taxonomy and sorption-relevant properties (FRÄNZLE & KUHNT 1994; KUHNT 1993; VETTER & KUHNT 1994).

h) World reference base for soil resources

The project of the International Reference Base for Soil Classification was taken up in 1982 as one of the proposed programmes to implement a World Soils Policy through UNEP. The German contributions to this projects are due to BLUME (Department of Soil Science, Kiel University) and BRONGER (Department of Geography, Kiel University). The objectives are:

– To develop an internationally acceptable framework for the description of soil resources to which national classifications can be attached and related, using FAO's Revised Legend as a guideline.
– To make this framework suitable for different applications in related fields such as agriculture, ecology, and environmental monitoring and assessment.

A world reference system for soil resources is a tool for the identification of pedological structures and the evaluation of their significance. It is intended to promote common understanding in soil science and to facilitate:
– the implementation of soil inventories and the transfer of pedological data
– the international use of pedological data for:
  · the evaluation of soil resources and their potential use patterns,
  · monitoring of soils, particularly soil development as related to land use,
  · the validation of experimental methods of soil use for sustainable development and
  · the transfer of soil-use technologies from one region to another (ISSS-ISRIC-FAO 1994).

References

DOWNING, J. A. (1991): Comparing apples with oranges. Methods of interecosystem comparison. In: COLE, J.; LOVETT, G.; FINDLAY, S. (eds.): Comparative analysis of ecosystems. Patterns, mechanisms, and theories. – New York (...), pp. 24–45
DUARTE, Ch. M. (1991): Variance and the description of nature. In: COLE, J.; LOVETT, G.; FINDLAY, S. (eds.): Comparative analysis of ecosystems. Patterns, mechanisms, and theories. – New York (...), pp. 301–318
FRÄNZLE, O. (1984): Regionally representative sampling. In: LEWIS, R.A.; STEIN, N.; LEWIS, C.W. (eds.): Environmental specimen banking and monitoring as related to banking. – Boston (...), pp. 164–179
FRÄNZLE, O. (1987): Regionally representative selection of soil samples for an environmental specimen bank with particular reference to loess soils. In: GeoJournal, 15.2, pp. 201–208
FRÄNZLE, O. (1990): Representative sampling of soils in the Federal Republic of Germany and the EC Countries. In: LIETH, H.; MARKERT, B. (eds.): Element concentration cadasters in ecosystems. – Weinheim, pp. 197–202
FRÄNZLE, O. (1994): Representative soil sampling. Problems and results in the development of international standards for sampling and pretreatment of soils. In: MARKERT, B. (ed.): Environmental sampling for trace analysis. – Weinheim, pp. 217–227
FRÄNZLE, O. KUHNT, G. (1994): Fundamentals of representative soil sampling. In: KUHNT, G.; MUNTAU, H. (eds.): EURO-Soils. Identification, collection, treatment, characterization. – Ispra, pp. 11–29
FRÄNZLE, D.; RUDOLPH, H.; DÖRRE, U. (1992): Erarbeitung und Erprobung einer Konzeption für die ökologisch orientierte Planung auf der Grundlage der regionalisierenden Umweltbeobachtung am Beispiel Schleswig-Holsteins. – Berlin (UBA-Texte 20/92)
GENTILE, S. H.; SLIMAK, M.W. (1992): Endpoints and indicators in ecotoxicological risk assessments. In: MCKENZIE, D. H.; HYATT, D. E.; MCDONALD, V.J. (eds.): Ecological indicators. – Vol. 2, London, New York, pp. 1385–1397
GÜNZLER, H. (1994): The accreditation of analytical laboratories. Quality assurance in Analytical Chemistry. In: Environmental Science & Pollution Research, 1 (1), pp. 56–57
ISSS-ISRIC-FAO (International Society of Soil Science – International Soil Refernce Information Centre – Food

and Agriculture Organization of the United Nations) (1994): World refernce base for soil resources. – Wageningen/Rome (Draft)

KUHNT, G. (1993): The Euro-soil concept as a basis for chemicals testing and pesticide research. In: MANSOUR, M. (ed.): Fate and prediction of environmental chemicals in soils, plants, and aquatic systems. – Boca Raton (...), pp. 83–93

LINTHURST, R. A.; THORNTON, K. W.; JACKSON, L. E. (1992): Integrated monitoring of ecological conditions: issues of scale, complexity, and future change. In: MCKENZIE, D. H.; HYATT, D. E.; MCDONALD, V. J. (eds.): Ecological indicators.- London, New York, Vol. 2, pp. 1421–1441

MEENTEMEYER, V.; BOX, E. O. (1987): Scale effects in landscape studies. In: TURNER, M. G. (ed.): Landscape heterogeneity and disturbance. – New York (...) (Ecological Studies, Vol. 64), pp. 15–34

MESSER, J. J. (1992): Indicators in regional ecological monitoring and risk assessement. In: MCKENZIE, D. H.; HYATT, D. E.; MCDONALD, V. J. (eds.): Ecological indicators. – London, New York, Vol. 1, pp. 135–146

MILNE, B. T. (1992): Indicators of landscape condition at many scales. In: MCKENZIE, D. H.; HYATT, D. E.; MCDONALD, V. J. (eds.): Ecological indicators. – London, New York, Vol. 2, pp. 883–895

REAGAN, D. P.; FORDHAM, C. L. (1992): An approach for selecting and using indicator species to monitor ecological effects resulting from chemical changes in soil and water. In: MCKENZIE, D. H.; HYATT, D. E.; MCDONALD, V. J. (eds.): Ecological indicators. – London, New York, Vol. 2, pp. 1319–1339

ROBERTSON, D.; BUNDY, A.; MUETZELFELDT, R.; HAGGITH, M.; USCHOLD, M. (1991): Eco-Logic. Logic-based approaches to ecological modelling. – Cambridge, London

SCHMIDT, R.; SCHRÖDER, W.; TAPKENHINRICHS, M. (1995): Soil data for models and resource management on different scales. In: Proceedings of the XVth World Congress of Soil Science in Acapulco, pp. 58–66

SCHRÖDER, W.; GARBE-SCHÖNBERG, C.-D.; FRÄNZLE, O. (1991): Die Validität von Umweltdaten. In: Umweltwissenschaften und Schadstoff-Forschung – Zeitschrift für Umweltchemie und Ökotoxikologie, 3 (4), S. 237–241

SCHRÖDER, W.; VETTER, L.; FRÄNZLE, O. (1992): Einfluß statistischer Verfahren auf die Bestimmung repräsentativer Standorte für Umweltuntersuchungen am Beispiel der neuen Bundesländer. In: Petermanns Geographische Mitteilungen, Jg. 136, H. 5/6, S. 309–318

VETTER, L. (1989): Evaluierung und Entwicklung statistischer Verfahren zur Auswahl von repräsentativen Untersuchungsobjekten für ökotoxikologische Problemstellungen. – Diss., Kiel

VETTER, L.; KUHNT, G. (1994): Methodological aspects of reference soil sampling. In: KUHNT, G.; MUNTAU, H. (eds.): EURO-Soils. Identification, collection, treatment, characterization. – Ispra, pp. 31 – 40

VETTER, L.; MAASS, R.; SCHRÖDER, W. (1991): Die Bedeutung der Repräsentanz für die Auswahl von Untersuchungsstandorten am Beispiel der Waldschadensforschung. In: Petermanns Geographische Mitteilungen, Jg. 136, H. 5/6, S. 165–175

WARD, R. C. (1992): Indicator selection: A key element in monitoring system design. In: MCKENZIE, D. H.; HYATT, D. E.; MCDONALD, V. J. (eds.): Ecological indicators. – London, New York, Vol. 1, pp. 147–157

2.3 Example for, and contribution to GTOS at global, regional and national level

2.3.1 Data management in ocean sciences

D. Kohnke

In oceanography there are operational programmes for collection, transmission, analysis and archival of data as well as scientific programmes for the investigation of ocean phenomena. Monitoring of oceanographic activities and of data exchange streams are fairly well organized.

Ocean data are collected in national or international scientific programmes, in monitoring programmes from moving and fixed platforms, and from satellites. Global scientific programmes are, for example, the "World Ocean Circulation Experiment" (WOCE), „Tropical Ocean and Global Atmosphere" (TOGA) and the "Joint Global Ocean Flux Study" (JGOFS).

There are monitoring programmes on a global as well as on a regional scale. The "ship-of-opportunity" programme of the Joint WMO[1]-IOC[2] Integrated Global Ocean Services System (IGOSS) provides temperature and salinity data from the upper 1000 m of the world oceans. These data are transmitted from the data collection platform to a land station via satellite in real time. Specialized oceanographic centres quality control the data and put it onto the Global Telecommunication System (GTS) of WMO in standardized formats for rapid transmission worldwide.

There are also unattended buoys either fixed at a certain position or drifting in the oceans which collect meteorological and oceanographic data. Their data are also transmitted via satellites in real time.

Within regional conventions for protecting the ocean from pollution, such as the Oslo, Paris or Helsinki Conventions, a great variety of variables is collected: chemical values, such as nutrients, heavy metals in suspended particles and in sediment, organic matters, radioactivity in water and in sediment; biological and chemical values in living resources. These data are exchanged in internationally agreed formats in non-real time.

The following mechanisms have been established within the international oceanographic data and information exchange (IODE) of the Intergovernmental Oceanographic Commission (IOC) of UNESCO to monitor the flow of data after their collection:

Publication of cruise schedules

Member States of the IOC publish the planned national oceanographic cruises about one year in advance. Annex A gives an example from the annual brochure "Fahrten deutscher Forschungsschiffe" (Cruise Schedule of German Research Vessels). This announcement gives administrators and data centres the very first information about possible data collection in oceanography. Particularly, national oceanographic data centres (NOD Cs) become aware of data which will finally be archived by them.

Cruise Summary Report

After the termination of a cruise the chief scientist must report to his NODC about the actual data collection which has taken place during his cruise. This report includes information about data types and the responsible scientist/institute/lab as well as time and area of data collection. Annex B shows the form for the „Cruise Summary Report". From now on the data centre is fully informed about the existence of data sets in various places in the country.

Data exchange

According to international rules data originators have to submit their data to their NODC within two years time. If they do not follow the rules the NODC contact them for data submission, because of the information from the Cruise Summary Report.

The main task of an NODC is to compile the data from national authorities and universities, to quality control and flag questionable data, to archive the data for national purposes, and to submit them to the appropriate World Data Centre (WDC) for Oceanography for final archival on a global scale.

Specialized centres have been established to handle data from either a geographical region or a special scientific programme, from special measuring techniques or to render particular services to the oceanographic community.

Data Centre Network

In oceanography there exist some types of data centres; all together form the international network of oceanographic data centres.

The National Oceanographic Data Centre (NODC) is responsible for the compilation of national data sets, for the quality control, for the transformation into internationally agreed exchange formats, and for the submission to a Responsible NODC or the World Data Centres (WDCs) for Oceanography.

A Responsible National Oceanographic Data Centre (RNODC) is an NODC which has accepted a special function, in addition to its national obligations. This function can include data responsibility for a geographical region, for a specific programme, for a special measuring method or parameter or for the preparation of special oceanographic products. The RNODCs work on behalf of the WDCs.

There are WDCs for Oceanography in Washington, Moscow and Beijing. Their main function is the final archival of ocean data. There are other WDCs, for example for Solar and Terrestrial Sciences or for Glaciology.

The Global Ocean Observing System (GOOS)

A tremendous challenge for the management of ocean data will be caused by the Global Ocean Observing System (GOOS) which is being planned by IOC, WMO, UNEP and ICSU[3]. A great variety of ocean data has to be handled on a local, regional or global scale in a way so that multidisciplinary asessment and forecasts are possible.

Objectives

„The objective of GOOS is to ensure by cooperative efforts of Member States global long-term, systematic observations of the World Ocean and to provide a mechanism and infrastructure for data and information to be made available to participating institutions and nations from coastal and oceanic regions, including enclosed and semi-enclosed seas, to help solve problems related to changes in regional and global environments on various time scales" [1].

GOOS will be developed on a sound scientific basis using the findings of existing on-going programmes, such as WOCE, TOGA and JGOFS. The implementation of GOOS will be based on existing operational programmes including IGOSS[4], IODE and GLOSS[5]. GOOS will utilize operational observing methods, both remote sensing and in situ measurements obtained from ships, towed and moored systems, drifting buoys and sub-surface floats.

The design of GOOS and the use of GOOS data need to be closely linked with the recent and continuing advances of numerical ocean and coupled ocean-atmosphere modelling. The concept for GOOS is based on the interdependence among observations, data assimilation and numerical models.

GOOS has been divided into different sets of aims and products to be achieved and produced; they constitute the five following GOOS modules:

1) Climate monitoring, assessment and prediction
2) Monitoring and assessment of marine living resources
3) Monitoring of the coastal zone environment and its changes
4) Assessment and prediction of the health of the ocean
5) Marine meteorological and oceanographic operational services

Module 1) is also the ocean component of the Global Climate Observing System (GCOS) which is being planned and developed by WMO.

Data management and data analysis

GOOS must be more than just a data collection programme. It must include assimilation, assessment and interpretation of the data as well as assuring a high data quality. The data must be made available in a timely fashion. Existing data management systems for real time, near-real time and in delayed mode will serve as a basis for the establishment of the GOOS data management system. But they need to be further improved to ensure an easy and quick access of GOOS users to relevant data bases.

To permit the detection of trends and of global changes continuous monitoring is needed using standard and consistent observation techniques to produce compatible data.

Additional national and international data assembly and analysis centres must be established for the preparation of local, regional and also global products. These centres must be equipped with [1]

a) advanced data base management and telecommunication systems capable of handling large volumes of incoming data and outgoing products;

b) data assimilation models able to generate complete four-dimensional data fields of known quality, i.e. with automatic quality control procedures;

c) means to provide timely and accurate data sets and the associated analytical procedures to less capable national centres on a quasi-real time mode.

Notes and References

1 World Meteorological Organization
2 Intergovermental Oceangraphic Commission
3 International Council of Scientific Unions
4 Integrated Global Ocean Services System (jointly run by IOC and WMO)
5 Global Sea-Level Observing System

[1] Anonymous: Global Ocean Observing System – Draft Development Plan; Document IOC/EC-XXV/8 Annex 1, 1992
[2] Anonymous: The Approach to GOOS; Document IOC-XVII/8 Annex 2 rev., 1993

Annex A:
Cruise Schedule of German Research Vessels

Schiffsname:
FS „METEOR"

Einsatzplanung: Institut für Meereskunde an der Universität Hamburg, Troplowitzstr. 7, 22529 Hamburg

Länge des Schiffes: 97,5 m; Breite des Schiffes: 16,5 m; Rufzeichen: DBBH;
Tonnage: 3990 BRT; Reisegeschwindigkeit: 12 Kn: Tiefgang: 5,6 m;

Fahrt-Nr.	Zeitraum	Arbeitsgebiet und wiss. Fragestellung	Untersuchungs-methoden	Fahrtleiter/Institut	Plätze f. Gastwiss.	Bemerkungen Ausgangs – Zielhafen
27/1	29.12.93–17.01.94	Biskaya OMEX		Balzer MCUHB	*)	Hamburg – La Coruna
27/2	20.01.94–15.02.94	NO-Atlantik BIO-FLUX		Pfannkuche GEOMAR	*)	La Coruna – Pt. Delgada
27/3	18.02.94–26.03.94	Trop. Atlantik WOCE	Cm, Do, O_2, Pro STD	Schott IFMKI	2	Pt. Delgada – Recife
28/1	29.03.94–11.05.94	Südatlantik WOCE	CTD, Rosette etc. Tracer	Müller IFMKI	*)	Recife – Pt. Noire
28/2	14.05.94–14.06.94	Südatlantik WOCE	CTD, Rosette etc. Tracer	Zenk IFMKI	*)	Pt. Noire – Buenos Aires
29/1	17.06.94–13.07.94	West.-Südatlantik	B, C, Ch, HyS, MB, Nt, PaS, Pr, Sd, Sei	Schulz GUHB	–	Buenos Aires – Montevideo
29/2	15.07.94–08.08.94	Westl.-Südatlantik	B, C, Ch, HyS, MB, Nt, PaS, Pr, Sd Sei	Bleil GUHB	–	Montevideo – Rio de Janeiro
29/3	11.08.94–05.09.94	Westl.-Südatlantik	B, C, Ch, HyS, MB, Nt, PaS, Pr, Sd Sei	Henrich GUHB	3–4	Rio de Janeiro – Las Palmas
30/1	07.09.94–20.09.94	Biskaya OMEX		Pfannkuche GEOMAR	*)	Las Palmas – Hamburg
30/2	12.10.94–13.11.94	Nordatlantik WOCE-Schnitt 48 °C	CTD-Wasserschöpfer BHS	Koltermann	*)	Hamburg – St. Johns
30/3	16.11.94–22.12.94	Labradorsee–WOCE-Schnitt 48 °N	CTD-Wasserschöpfer	Meincke IFMHH	2	St. Johns – Hamburg

*) Nach Rücksprache mit dem Fahrtleiter

Spezielle Angaben über die wissenschaftlichen Arbeiten:

Att	–	Attenuation
Att (a)	–	Attenuation
B	–	Bottom sample
Ba	–	Bathymetry
Bac	–	Bacteriology
Bd	–	Biological dredge
Be	–	Benthos
Bf	–	Bottom photography
Bo	–	Botany
Bt	–	Bathythermograph or X-BT
C	–	Coring
Ch	–	Chemistry
Ch (air)	–	Air chemistry
CIE	–	Chromyticity coordinates
D	–	Dredging
Dd	–	Dye diffusion
Dv	–	Diving
F	–	Fish sampling
Fm	–	Fish migration
Ge	–	Gear experiment
Gm	–	Geomorphology
Gr	–	Grab
Grav	–	Gravity
Gs	–	Coastal geodetic survey
HF	–	Heat flow
Hm	–	Heavy metals
HyS	–	Hydrosweep
Inst	–	Instrument development
Irr	–	Irradiance
MB	–	Microbiology
MCS	–	Multichannel seismic reflection
Mg	–	Geomagnetic
N	–	Neuston
Nt	–	Nutrients
O_2	–	Oxygen
P	–	Pollution
Pa	–	Paleontology
PaS	–	Parasound
Pr	–	Primary production
Pro	–	Current profiler
Qm	–	Quantameter
ROV	–	Remote Operating Vehicle
Sa	–	Salinity
SCS	–	Single channel seismic reflection
Sd	–	Sedimentology
Se	–	Seston
Sei	–	Seismology
Sm	–	Suspended matter
SR	–	Seismic refraction
STD	–	Salinity/Temperature/Depth device
Sv	–	Sound velocity
T	–	Temperature
Ta	–	Tagging
Th	–	Thermistor chain
Ti	–	Tides
Tr	–	Transparency
Tur	–	Turbidity
TV	–	Underwater television
U	–	Underwater acoustics
V	–	Visibility
W	–	Waves
Xt	–	Sub-surface tow
YS	–	Yellow Substance
Z	–	Zoology

Annex B: Cruise Summary Report

CRUISE SUMMARY REPORT

FOR COLLATING CENTRE USE

Centre: .... Ref. No: ....

Is data exchange restricted? ☐ Yes ☐ In part ☐ No

SHIP enter the full name and international radio call sign of the ship from which the data were collected, and indicate the type of ship, for example, research ship; ship of opportunity, naval survey vessel; etc.

Name: .... Call Sign: ....

Type of ship: ....

CRUISE NO./NAME .... enter the unique number, name or acronym assigned to the cruise (or cruise leg, if appropriate).

CRUISE PERIOD start (set sail) |__|__| |__|__| |__|__|__|__| to |__|__| |__|__| |__|__|__|__| end (return to port)
                                   day   month   year   day   month   year

PORT OF DEPARTURE (enter name and country) ....

PORT OF RETURN (enter name and country) ....

RESPONSIBLE LABORATORY enter name and address of the laboratory responsible for coordinating the scientific planning of the cruise.

Name: ....

Address: ....

Country: ....

CHIEF SCIENTIST(S) enter name and laboratory of the person(s) in charge of the scientific work (chief of mission) during the cruise.

OBJECTIVES AND BRIEF NARRATIVE OF CRUISE enter sufficient information about the purpose and nature of the cruise so as to provide the context in which the reported data were collected.

PROJECT (IF APPLICABLE) if the cruise is designated as part of a larger scale cooperative project (or expedition or programme), then enter the name of the project, and of the organisation responsible for coordinating the project.

Project name: ....

Coordinating body: ....

PRINCIPAL INVESTIGATORS: Enter the name and address of the Principal Investigators responsible for the data collected on the cruise, and who may be contacted for further information about the data. (The letter assigned below against each Principal Investigator is used on pages 2 and 3, under the column heading 'PI', to identify the data sets for which he/she is responsible)

A.

B.

C.

D.

E.

F.

MOORINGS, BOTTOM MOUNTED GEAR AND DRIFTING SYSTEMS

This section should be used for reporting moorings, bottom mounted gear and drifting systems (both surface and deep) deployed and/or recovered during the cruise. Separate entries should be made for each location (only deployment positions need be given for drifting systems). This section may also be used to report data collected at fixed locations which are returned to routinely in order to construct 'long time series'.

PI see top of page.	APPROXIMATE POSITION		DATA TYPE enter code(s) from list on cover page.	DESCRIPTION Identify, as appropriate, the nature of the instrumentation, the parameters (to be) measured, the number of instruments and their depths, whether deployed and/or recovered, dates of deployment and/or recovery, and any identifiers given to the site.
	LATITUDE deg min N/S	LONGITUDE deg min E/W		

Please continue on separate sheet if necessary.

SUMMARY OF MEASUREMENTS AND SAMPLES TAKEN

Except for the data already described on page 2 under 'Moorings, Bottom Mounted Gear and Drifting Systems', this section should include a summary of all data collected on the cruise, whether they be measurements (e.g. temperature, salinity values) or samples (e.g. cores, net hauls).

Separate entries should be made for each distinct and coherent set of measurements or samples. Different modes of data collection (e.g. vertical profiles as opposed to underway measurements) should be clearly distinguished, as should measurement/sampling techniques that imply distinctly different accuracies or spatial/temporal resolutions. Thus, for example, separate entries would be created for i) BT drops, ii) water bottle stations, iii) CTD casts, iv) towed CTD, v) towed undulating CTD profiler, vi) surface water intake measurements, etc.

Each data set entry should start on a new line - it's description may extend over several lines if necessary.

NO, UNITS : for each data set, enter the estimated amount of data collected expressed in terms of the number of: 'stations'; 'miles' of track; 'days' of recording; 'cores' taken; net 'hauls'; balloon 'ascents'; or whatever unit is most appropriate to the data. The amount should be entered under 'NO' and the counting unit should be identified in plain text under 'UNITS'.

PI	NO	UNITS	DATA TYPE	DESCRIPTION
see page 2	see above	see above	enter code(s) from list on cover page.	Identify, as appropriate, the nature of the data and of the instrumentation/sampling gear and list the parameters measured. Include any supplementary information that may be appropriate, e.g. vertical or horizontal profiles, depth horizons, continuous recording or discrete samples, etc. For samples taken for later analysis on shore, an indication should be given of the type of analysis planned, i.e. the purpose for which the samples were taken.

Please continue on separate sheet if necessary.

TRACK CHART: You are strongly encouraged to submit, with the completed report, an annotated track chart illustrating the route followed and the points where measurements were taken.

Insert a tick (✓) in this box if a track chart is supplied. ☐

GENERAL OCEAN AREA(S): Enter the names of the oceans and/or seas in which data were collected during the cruise - please use commonly recognised names (see, for example, International Hydrographic Bureau Special Publication No. 23, 'Limits of Oceans and Seas').

....
....

SPECIFIC AREAS: If the cruise activities were concentrated in a specific area(s) of an ocean or sea, then enter a description of the area(s). Such descriptions may include references to local geographic areas, to sea floor features, or to geographic coordinates.

....
....
....

GEOGRAPHIC COVERAGE - INSERT 'X' IN EACH SQUARE IN WHICH DATA WERE COLLECTED

THANK YOU FOR YOUR COOPERATION

Please send your completed report without delay to the collating centre indicated on the cover page

2.3.2 Quality assurance/quality control as prerequisite for harmonization of data collection and interpretation within the WMO-Global Atmosphere Watch (GAW) Programme

V. A. Mohnen

The issues of global change and man's influence on the quality of the natural environment are complex. The chemistry of the atmosphere is changing: a global warming is expected due to increasing concentrations of greenhouse gases (CH_4, CO_2, N_2O, CFCs and ozone) and altered amounts of clouds and particles in the atmosphere; the stratospheric ozone layer and consequently the surface solar flux of ultraviolet radiation are being modified; the integrated oxidation rage, loosely referred to as the oxidation efficiency or oxidation capacity of the atmosphere is changing and the abundance of trace gases including those with significant greenhouse warming potential are perturbed. Assessment of the problems, their impacts and the responses to them occupy a prominent position of the international agenda today. Finding an ecologically effective and economically efficient solution to these problems confronts all countries of the world with a challenge of global dimensions.

The scientific input to the debate on environmental issues must derive from an adequate knowledge base. This can only be achieved through high quality, strategically oriented observations, and research related to the particular issues. This necessitates the establishment of proper global environmental observation systems. It is the only effective way to ensure systematic gathering of data world-wide according to comparable and clearly defined measuring criteria, to enable coordinated data processing and quality assurance, and to facilitate the distribution and provision of available information to the widely varied group of users. This complex international task must be tackled jointly by international organizations and the scientific community.

The Global Atmosphere Watch (GAW) program of WMO serves this purpose. It is an integral part of the Global Climate Observing System (GCOS) established by WMO, UNESCO/IOC, UNEP and ICSU. GAW is devoted to the investigation of changing chemical composition and related physical characteristics of the global atmosphere. GAW is a coordinated network of observing stations, associated facilities and infrastructure encompassing measurement and related scientific assessment activities. The overall role of GAW is to supply basic information of known quality indicative of the atmospheric environment that transcends specific issues.

The GAW measurement responsibilities include:
- greenhouse gases
- ozone (surface, total and profile)
- radiation (including UV-B) and optical depth
- precipitation chemistry
- chemical and physical properties of aerosols
- reactive gases
- radionuclides
- related meteorological parameters.

The data obtained by GAW initially provide:
- integrated monitoring

- scientific assessments
- early warnings.

The GAW-structure calls for (1) up to 30 global stations located at remote pristine locations; (2) over 100 regional stations for characterizing the regional environmental quality away from direct pollution sources; and (3) a mechanism/system of activities for producing measurements of known quality and of value for making environmental policy decisions (quality assurance/quality control).

WMO's strategy for implementing total quality assurance is built around Quality Assurance/Science Activity Centres (QA/SACs); following the recommendations of a meeting of experts assembled by WMO (March 26–30, 1992, Garmisch-Partenkirchen, Germany) and charged with the development of a quality assurance plan for GAW (WMO/GAW Report No. 80, 1992: „Report of the WMO Meeting of Experts on the Quality Assurance Plan for the Global Atmosphere Watch"). In June of 1992, the Executive Council of WMO, recalling previous concerns of data reliability and recognizing the importance of quality assured data, fully supported the actions taken by the Secretary-General and the EC Panel of Experts/CAS Working Group and endorsed the structure and establishment of QA/SACs within GAW. The Executive Council requested that the Secretary-General proceed with this development. Accordingly, the WMO arranged a second meeting of experts (December 7–11, 1992, Garmisch-Partenkirchen, Germany) to develop guidelines for the implementation of these QA/SACs (WMO/GAW Report No. 92, 1993 "Report of the WMO Meeting of Experts on the Implementation of Quality Assurance/Science Activity Centres for GAW").

The essential functions of these QA/SACs include the preparation and execution of (1) the comprehensive GAW Quality Assurance Programme for all measurement components, and (2) a plan for education and capacity building within the countries that have committed to maintain and operate GAW sites. Building the indigenous technical and scientific capabilities and infrastructure (particularly in developing countries) is an integral part of an effective QA programme.

The EC Panel of Experts/CAS Working Group on Environmental Pollution and Atmospheric Chemistry (The Panel) has reviewed the WMO Reports No. 92 and 93 and, in its role as an advisory body, endorsed its recommendations. The Panel reaffirmed the QA/SAC's role as a key element of the GAW system, and the members expressed their enthusiasm to contribute to their development and to provide whatever advice possible.

In the discussion of the priority for GAW measurements the Panel advised QA/SACs to focus attention initially on:

- Ozone (surface, total and vertical profile)
- Aerosols (implementation of the recommended GAW aerosol programme as defined in GAW Reports Nos. 79 and 80, especially regarding particle size spectrum and chemical composition, optical depth and condensation nuclei concentration was considered a high priority)
- UV-B radiation (considering the difficulties involved in measuring UV-B, particularly at low solar elevations, global stations are encouraged to measure UV-B with a spectral resolution of about 0.5 nm. Measurement of cloud cover, total and surface ozone, direct and diffuse solar radiation, optical depth and broad band UV-B should be made concurrently.)
- Precipitation (weekly samples, wet only)

The spread of QA/SAC activities will be broad, but first steps in their implementation need to be carefully directed. Contemporary concerns over possible chemical and climatic effects of the observed increases and decreases, respectively, in tropospheric and stratospheric ozone have resulted in selection of this issue as an example for describing the first detailed QA/SAC activity. A Quality Assurance Project Plan (QAPjP) has been prepared for surface ozone measurements which is now published by WMO and distributed to all GAW stations for implementation. A similar document for free tropospheric ozone – mainly ozone sonde – measurements is in preparation. Eventually, there will be QAPjPs and associated standard operating procedures (SOPs) for each measurement parameter in GAW.

Functions of a QA/SAC

The overall function of the Quality Assurance/Science Activity Centres (QA/SACs) has been discussed in detail in WMO/GAW Report Nos. 80, 92, and 93. The QA/SACs provide to WMO the communications and constructive assistance necessary for maintaining consistent and known data quality in the Global Atmosphere Watch Programme so that it can achieve its monitoring goals. They are an essential and integral part of the GAW network and the operational mechanism for harmonization and coordination of GAW activities. All GAW participants will abide by the WMO policies and the QA/SAC procedures and methodologies to produce data bases of consistent and known quality. Although the country supporting the GAW site remains the primary owner of its site's data and may release it as agreed with WMO, the data can only be accepted as GAW data after completing the QA/QC procedures for GAW.

The QA/SACs will facilitate transfer of knowledge, experience and expertise within the network to ensure common practices. While the QA/SACs perform a network-wide quality review, the site Principal Scientist (PI) has primary responsibility for the quality of the data generated at his/her site. The QA/SACs will help establish and foster research ties among site PIs and with users of GAW data bases to ensure relevance to international scientific and policy concerns.

The QA/SACs will be advised by a policy advisory panel of neutral experts and by technical advisory panels which include site PIs. The effectiveness of the QA programme and QA/SAC procedures will be reviewed by WMO on a regular basis, e.g., after 3-5 years of operation.

Specific Duties of QA/SACs

In the following, the word "data" is taken to mean both measurements of primary environmental variables and necessary support variables (metadata). The specific duties include:

- Management of the preparation of the QA plans for environmental variables
  - ensure the production of specifications to Data Quality Objectives (DQOs), Quality Assurance Project Plans (QAPjPs), lists of supporting information to be supplied and the corresponding protocols for flags
  - perform pilot studies to test QC procedures on a limited scale before full implementation
  - coordinate activities among all QA/SACs

- Quality assurance support
  - design and lead QC tests in consultation with sites to resolve performance questions

- prepare (or identify existing) QA/QC and data transfer protocols (For many environmental variables, QA procedures are already well established, and recognized authorities exist within the regions who are likely to be willing consultants.)
  - prepare QC activity schedule; site-visits, expert consultations (performance and systems audits), calibrations, and intercomparisons
  - promote twinning arrangements between sites
  - identify sources of standards and instructions for usage (provide if necessary)
  - periodically review site adequacy (in specific situations, QA activities may be delegated to appropriate expert groups outside the GAW network)

- Management Review of Data and QC Products
  - perform quality checks on data streams and QA products (primary quality checks must be performed by the site PI)
  - transmit data, metadata, and flags to data centre following review within the specified period
  - if review has not occurred within specified time, transmit data, metadata, and flags to Data Centre with a flag indicating that review has been delayed
  - until QA procedures are fully implemented, all data, metadata and flags will pass through the QA/SAC with flags where appropriate, indicating that the QA/SAC has not reviewed a part of the data because the QA procedure for that part has not been elaborated.
  - allocate station performance categories

- Training
  - develop an ongoing, network-specific educational programme (training courses, seminars, etc.)
  - prepare development plans for technical personnel (temporary appointments, etc.) and certification procedures to ensure that only trained personnel are involved in station operations

- Communication and Dissemination of Information
  - establish a communications network (e.g. e-mail, newsletter, bulletin boards, clearing house for urgent operational and other information, etc.)
  - promote awareness of technological changes (workshops, training, information),
  - arrange periodic review meetings (including scientists, technicians and operators from GAW and other scientists whose work is of relevance to GAW) to present scientific results, exchange technical information, and facilitate coordination of QA procedures,
  - cooperate with international scientific programs involved in atmospheric research such as IGBP-IGAC.

- Duties of Data Centres and Associated QA Officer
  - collect the entire data set (with flags), specified data documentation and QA reports for one or more specific components
  - assess QA reviews and data flags from QA/SACs to ensure uniform performance across the regions
  - send assessment reports to WMO copied to QA/SACs and respective GAW sites
  - final acceptance of flag designations
  - merge data and flags from QA/SACs
  - store data base to the DC for permanent storage and public distribution.

It is assumed that the data centres, in cooperation with QA/SACs, will establish a system to ensure that the entire data set is harmonized.

- Linkages and Information Flows. The flows of data, metadata, and flags, as well as the linkages between the sites, SAC, and data centres are shown in Figure 2.3.2.1.

- Paradigm for Quality Assurance and Measurement-System Development. Figure 2.3.2.2 provides a pictorial description of the general approach to be applied for quality-assurance and measurement system development. Here it is important to note the hierarchical decendency of elements from top to bottom, with each function fulfilling the needs of the function immediately above. Thus the scientific objectives associated with the measurement of a given parameter (Box 2) are chosen to resolve the issues identified by the GAW administration in consultation with The Panel. These scientific objectives are chosen so as to most effectively contribute to resolution of these issues within the practical constraints of current scientific knowledge and resource limitations. Initial assignments of required measurement uncertainty levels are established at this point to guide operations in succeeding boxes.

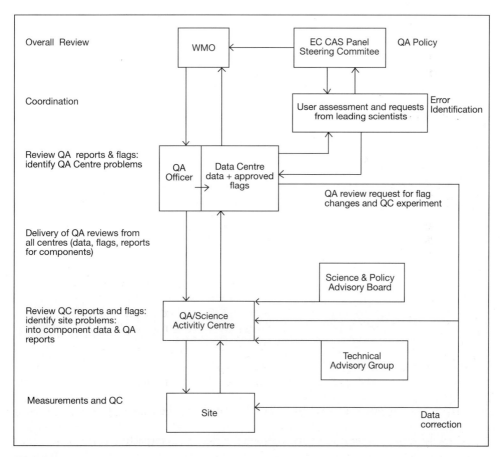

Fig. 2.3.2.1:
Flow Diagram of Quality Assurance (QA) Implementation Plan for GAW

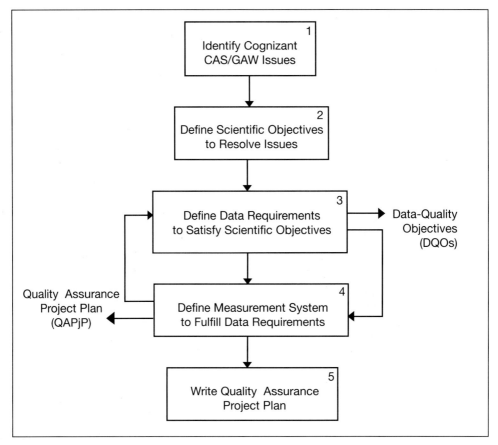

Fig. 2.3.2.2:
Road Map for Quality Assurance and Measurements-System Development

Box 3–4 addresses design of the scientific measurement system to fulfill the scientific objectives, in combination with the quality-assurance program associated with these measurements. These are closely coupled because their design is necessarily a joint, iterative process: the initial scientific design is challenged by the initial data-quality requirements, with resulting modifications, and the process continues until an acceptable convergence is attained. Typically the scientific design is conducted by posing a series of specific questions regarding the appropriateness of the measurement system to satisfying the scientific objectives. These usually include (but are not restricted to) questions pertaining to required spatial resolution, temporal resolution, instrument accuracy, redundancy, and ancillary measurements.

The QA/SAC activity described in this hierarchical cascade is intended for promoting group input to the design process within a workshop forum. It parallels the processes for formally establishing the Data Quality Objectives and QA Project Plans – functions which are normally carried out by individuals or smaller teams on a more intensive basis. As such, this activity is intended as a group launching operation for DQO and QAPjP formulation; if conducted with appropriate guidance it should provide much of the information required for these later processes.

Training is required for three categories of station personnel: the scientific manager responsible for the overall operation of a GAW station, the scientist responsible for component measurement programs, and the technician responsible for operating the measurement systems.

The scientist who has overall responsibility for the management and operation of a GAW global station must be familiar with the scope and goals of the GAW, with the role of his station within GAW, with the detailed operation of at least one (usually one of the more complex) measurement program, and with all other components programs and quality assurance and quality control (QA/QC) procedures at the station.

The scientist running an individual program must be fully acquainted with all aspects of his and related measurements, and with similar programs at other GAW stations. Technical staff require detailed training in the measurements systems they must operate. In all cases there is a need for overlapping training so that measurements are not jeopardized during the absence of the primary scientist or technician.

The training program of QA/SACs for new GAW global stations will include the following elements:

1. Initial short courses or training workshops on all aspects of the operation of a GAW station and on the measurement programs to be carried out at the new station. This will include introductory material on GAW, instruction on the rationale for and theory of the measurements, hands-on familiarization with instruments and procedures, familiarization with QA/QC procedures, and instruction on analysis and use of the data generated. This would be an intensive one- to three-week course at a fairly high level, to be attended by scientists and technicians. Training workshops are planned for Greece in October 1993 and for Argentina in May 1994. The latter should serve as a pilot workshop for the development of a comprehensive training program for the QA/SAC-Americas.

2. Carefully designed training programs are a second element. Technicians (and perhaps young scientists) will undergo a structured training program at an established GAW global station or associated laboratory, at a QA/SAC, for a period of several weeks to three months. The trainee will be instructed by a mentor, or series of mentors, in all relevant measurement programs and QA/QC procedures, and will be tested at the completion of his training period. Training opportunities of this type should be arranged through the QA/SACs at a number of established GAW global stations or their associated laboratories (Cape Grim; Mauna Loa Observatory; CMDL, Boulder CO USA; AES, Toronto, Canada; IFU, Garmisch-Partenkirchen Germany). Regular re-certification would be an element of this type of training. A significant advantage of this component of training is that valuable contacts are established between the new and the established station.

3. Expert visits are essential to assist with the set-up and initial operation of new stations. One or more experts may be involved in visits lasting from two weeks to two or three months in total. The experts provide hands-on assistance with measurement systems, QA/QC procedures and trouble-shooting. Follow-up visits of a similar nature, but shorter duration, are recommended.

4. Occasional short courses on specific measurement topics may be necessary, for example, when a new measurement program or a new instrument is added. An example here is a planned one-to two-week training course in China on the operation of the Brewer spectrophotometer.

5. Similarly, occasional seminars on selected topics may be offered by visiting experts. These can serve to introduce new methods or instruments, to reinforce earlier training, or to provide an opportunity for upgrading of technical staff.

6. It is anticipated that the QA/SACs will communicate with the stations on a regular basis, by electronic mail-bulletin boards, or a newsletter. Such communication is important to disseminate advisory notices on measurement programs, to announce training seminars and conferences, and to provide feedback to station operators on QA/QC exercises and on the use of data collected at the station.

7. An effective means of transmitting information on instrument operation and measurement techniques in with videos, prepared either by experienced station operators or by instrument manufacturers. This is often a cost-effective means to bring operators up to date and to provide ready backup on repairs and other technical matters.

8. An essential continuing aspect of training and education for station scientists and senior technicians is attendance at regular (annual) meetings of global station personnel. Such meetings are held annually for the U.S. program by CMDL, and on a regular basis for the Cape Grim Observatory. These provide an opportunity to exchange experiences and information, to establish contacts with colleagues at other stations, and to disseminate recent results. Initially, the possibility of convening such a meeting of new GAW station scientists in association with the American or Australian meetings should be explored.

9. A requirement of the QA/SAC program is that stations undergo a regular systems audit. In addition to checking station operations, this provides an opportunity for feedback to station personnel on their performance and on potential improvements to be made. The QA/SACs are expected to be in regular direct contact with station personnel, for problem solving and encouragement.

10. A training mechanism used by the Fraunhofer Institute that has the potential for application to the new GAW stations was noted. Complete station measurement systems are assembled in a container. This is used as a training module for new station operators initially at the Institute, and then it is shipped for installation at a new measurement site. Here operations can begin quickly by the operators who are already familiar with the station.

Conclusion

The scientific input to the debate on environmental issues must derive from an adequate knowledge base. This can only be achieved through high quality, strategically oriented observations, and research related to the particular issues. This necessitates the establishment of proper global environmental observation systems. It is the only effective way to ensure systematic gathering of data world-wide according to comparable and clearly defined measuring criteria, to enable coordinated data processing and quality assurance, and to facilitate the distribution and provision of available information to the widely varied group of users. This complex international task must be tackled jointly by international organizations and the scientific community.

2.3.3 Terrestrial monitoring in the Arctic environment

L.-O. Reiersen

Introduction

Increased awareness of anthropogenic pollution in the Arctic led to the adoption of the Arctic Environmental Protection Strategy at the Ministerial meeting of the circumpolar nations in Finland,. June 1991. One of the key aims of the strategy is to examine the problem of pollution in this area, to achieve this the Arctic Monitoring and Assessment Program (AMAP) was established. The main objective of AMAP is to monitor the levels of pollutants and assess the effects of anthropogenic pollution in all components of the Arctic environment. A Task Force (AMATF) was established to implement AMAP. This group was substituted by a Working Group (AMAP WG) in 1993. The AMAP WG is responsible for preparation of the first holistic status report, which will be presented to the Ministers of the eight Arctic countries in 1996 (for more details on AMAP see Annex II-3.4).

The terrestrial environment is one of the five major components of the Arctic environment. An understanding of the patterns and effects of contaminants in the terrestrial environment is important not only for monitoring and assessing the terrestrial environment, but also for assessing other components of the Arctic environment. Contaminants deposited in the terrestrial environment can contaminate other elements of the Arctic ecosystem, i.e. fresh and marine waters. The terrestrial environment also contains a food web important to indigenous Arctic peoples. Finally, the terrestrial environment offers opportunities in access and in stability that make attribution of contaminants to sources (discharges to air, water or land) more tractable than is the case in other environments.

Specific objectives for AMAP's terrestrial sub-programme

i) Describe the broad patterns of contamination in terrestrial ecosystems.
    a) Describe current patterns and concentrations
    b) Quantify historical contamination
    c) Separate anthropogenic contributions from „natural" contributions (for heavy metals and radionuclides)
ii) Identify locations of higher potential exposure of human populations to anthropogenic contaminants.

Priority interactions for the terrestrial sub-programme with other AMAP sub-programmes

i) Atmospheric: Relate terrestrial contaminant patterns and history to global emissions sources through knowledge of transport models.
ii) Human Health: Relate human contaminants exposure or risk to food web bioaccumulation and spatial patterns.
iii) Freshwater: Determine the history of deposition in lake cores to partition anthropogenic from natural contributions.

Area description

Environmental monitoring should, where possible, take place in small catchment areas where a number of variables can be measured simultaneously. Suitable catchment areas do not, of course, occur in all areas for which monitoring is desirable (especially in the High Arctic), and therefore, alternate sites will have to be selected.

The geographical boundaries of Arctic and sub-Arctic regions may vary among countries depending upon the extent of the Arctic and sub-Arctic characteristics. Some countries may choose to include only sites situated north of the Arctic Circle whereas other countries may choose to extend their area of interest south to a latitude of 60°N or even further in certain instances; i.e. inclusion of northern Quebec and Labrador in Canada, and the Aleutian Islands of Alaska, USA.

Methodology

For the first phase of AMAP three categories of monitoring priorities have been recommended:
i) E – Essential to address the circumpolar assessment
ii) ES – Essential to address specific sub-regional assessment
iii) R – Recommended to provide more complete scientific substantiation to the circum polar and sub-regional assessment.

Within a few areas there exist ongoing monitoring activities and AMAP will build on these. However, for most areas in the Arctic there are no programmes or data available from research projects and in the first phase of AMAP „screening" or „baseline" monitoring will be performed Routine screening surveys provide information from all of the participating countries for circumpolar mapping exercises that describe the extent of a certain aspect of environmental monitoring, i.e. heavy metals in mosses. Baseline monitoring refers to the assemblage of data (existing or to be collected) in order to provide a status level of a given parameter against which to compare any data that may be collected in the future (i.e. data collected in relation to a contamination episode). In the event that recent baseline data do not already exist, or that there are gaps in the baseline data, each participating country will undertake efforts to obtain the recommended baseline data or to fill in the data gaps.

A specific list of methodology and quality assurance has been recommended for a number of components, including recommendations for the seasonal timing of collection and the appropriate part of tissue to be collected for sampling, storage, analyses etc. However, there is still lack of internationally recommended methods for some components, e.g. persistent organics, and some adjustments for Arctic climate and conditions will have to be made. Regarding biological effect studies in the Arctic there is a need to develop methods, and a specific workshop regarding terrestrial effect methods will be arranged during the winter of 1995.

Rationale for terrestrial media and parameters chosen for monitoring

Six specific pollution issues have been identified as requiring attention in the AMAP, they are issues associated with persistent organic compounds, heavy metals, radioactivity, oil, acidification and radiatively important trace species (RITS) e.g. greenhouse gases. Of these six issues, it was agreed that immediate attention would be focused primarily on three parameters: persistent organic compounds, heavy metals and radioactivity, and acidification in a sub-regional context.

The terrestrial media groups chosen were: soil, humus, lichens, mosses, mushrooms, higher plants, birds and mammals. These media were chosen to:

i) map contaminant levels (present, future and indications of historical contaminant levels),
ii) monitor contaminant levels in food webs; i.e. lichen-caribou/reindeer-wolf/man, and
iii) monitor most compartments of the terrestrial environment.
iv) biological effect studies.

The parameters that are recommended for monitoring in the various media have been specified.

Soil and Soilwater
Selected heavy metals, radionuclides, persistent organic compounds, acidification and various aspects of soil organisms. Measuring the levels of heavy metals in soil was considered essential (E) as one needs to calibrate/correct for the levels found in higher plants. Levels in soils might also explain contaminant levels in vegetation in areas where you have erosion/solifluction (dust spray). Organochlorines and radionuclides (at least cesium) are readily bound to colloids and not known to be taken up by plants in great quantity, therefore essential sub-regional (ES) sampling was recommended.

Humus
Selected heavy metals and persistent organic compounds. Measuring levels of heavy metals and organochlorines in humus, the „O" horizon (LFH), will add to the understanding of contaminants available to the higher plants from the whole plant substrate.

Lichens and Mosses
Selected heavy metals, persistent organic compounds and radionuclides. Both lichens and mosses are good indicators of atmospheric deposition of contaminants because they do not draw nutrients from the soil. Several species of lichens and mosses are circumpolar and therefore will provide important background information on contaminants reaching the terrestrial environment. Lichens and mosses also provide an important food base for several Arctic animals. Pleurozium and Hylocomium are predominantly forest species of mosses and therefore could be monitored in the low Arctic. Racomitrium, however, is a dry rock species and could be suitable for the higher Arctic region. There is a question as to whether these two species are suitable for the High Arctic (growth form, etc.). Monitoring of metals is recommended to be essential (E) in mosses and recommended (R) in lichens, except for areas where it is important as a part of good web studies, then it is essential (ES). Monitoring of organochlorines in both mosses and lichens is recommended (R). Measurements of radionuclides in mosses are recommended (R), but should be banked in order to have reference material in case of future episodes/incidents.

Mushrooms
Selected radionuclides and heavy metals. Mushrooms are susceptible to heavy metal and radionuclide contamination. In areas where there is concern with respect to their consumption by people and animals they should be monitored. The species of mushroom to be monitored should be selected accordingly.

Higher Plants
Selected heavy metals. Higher plants should be measured for heavy metals because metals are taken up by higher plants, and higher plants play an important role in the Arctic food chain.

Organochlorines and radionuclides will not be monitored because they are not readily taken up by higher plants. Species of higher plants and metals should be selected according to their importance in the regional food webs.

Birds and Mammals
Selected heavy metals and organic compounds.
Birds and mammals are chosen according to their position in the food chain and their role in the diet of northern native people.

i) *Birds*
Selected heavy metals and persistent organic compounds. Monitoring of heavy metals and organochlorines in birds is essential in special sub-regions in order to gather information on bioaccumulation of contaminants in the avian food web. It is suggested that rock ptarmigan and willow ptarmigan (both herbivorous species) and a predatory species such as snowy owl, gyrfalcon or merlin be monitored for heavy metals. However it is recommended that only the predatory species be monitored for the organochlorines, since organic compounds are not readily taken up in plants and therefore would not be expected to bioaccumulate in herbivorous species.

ii) *Mammals*
Selected heavy metals, radionuclides and persistent organic compounds. Due to the importance of caribou/reindeer (Rangifer sp.) in the diets of northern native people it is recommended that the collection of baseline data for heavy metals and radionuclides be essential. Due to the herbivorous habit of caribou/reindeer (problems with the analysis of persistent organic compounds in plants) measurements of organochlorines will be essential (E). Monitoring of heavy metals, persistent organic compounds and radionuclides is optional in special areas for the Arctic fox (Renard articque) since it is omnivorous and therefore integrates food web components of both the terrestrial and the marine environment. Monitoring of the same contaminants is also optional in the wolf (Canis lupus) in special areas since the wolf represents the top predator in a Arctic terrestrial food web.

iii) *Endangered Species*
Some of the recommended species of Arctic birds and mammals are listed as endangered in some areas of the Arctic. Information obtained from monitoring Arctic food chains may help to evaluate whether the endangered species are at risk due to contamination. In areas where these species are considered endangered, only hair from mammals, and eggs and feathers from birds will be sampled in order to avoid killing these animals.

Other issues

Monitoring Biological Effects
Only limited information is available on the relationship between pollution and effects on individual animals or animal populations. Further Arctic monitoring should also include monitoring for unknown pollution effects. Consequently, after the initial monitoring phase (1993-1996), monitoring of pollution levels should be supplemented with biological monitoring.

Biomarker techniques should be incorporated into future monitoring programs to screen wildlife for biological effects caused by elevated levels of contaminants. Biomarkers may be useful in determining the geographical distribution of chemical effects in the environment and their changes over time as well as determining the identity and source of chemical pollutants and establishing cause-effect linkages. In theory, all taxa of plants and animals may be useful, but in practice, species will have to be carefully selected based on knowledge of the necessary background information required for interpretation of results of the biomarker studies.

Changes in population size, age and sex composition, and reproductive success should be monitored in order to provide baseline data against which to evaluate possible impacts on human populations caused by contamination. Northern peoples may help to provide some of this information.

Acidification
Acidification due to anthropogenic pollution is a problem in certain regions of the Arctic. Consequently, parameters to monitor acidification in the soil are essential in special subregions, basically in forested areas in the Low Arctic. Due to the thin soil layers in the High Arctic regions, surface waters might for those areas be considered a more sensitive indicator of acidification effects than soil.

Climate change
Changes of vegetation or treeline borders due to climate change should be monitored using remote sensing techniques. Other techniques of monitoring climate change are covered in separate programmes.

2.3.4 Integrated environmental monitoring network development: The experience of the Environmental Change Network (ECN) (United Kingdom)

A. M. Lane

The ECN grew from proposals made by the UK House of Lords Report on Agricultural and Environmental Research in 1988, and from ideas within the UK Institute of Terrestrial Ecology (ITE) and other agencies on the need for dependable long-term data on important environmental variables. An integrated multi-sectoral approach to long-term environmental monitoring was seen as a fundamental basis for understanding the processes which drive and respond to change. Subsequent planning committees recommended that a network of representative sites be selected to monitor a range of physical, biological and chemical variables according to standard protocols (Sykes, 1994). A list of 24 possible terrestrial sites was compiled, selected according to the following criteria:

i) Good geographical distribution, and representing a range of environmental variation, including the principal semi-natural and managed ecosystems
ii) Some guarantee of long-term physical and financial security.
iii) A known history of consistent management
iv) Reliable and accessible past data, preferably at least 10 years
v) Sufficient size and opportunity to allow further experimentation and monitoring

ECN's approach was to begin monitoring with 8 'founding' terrestrial sites, and gradually to recruit more sites, including freshwater sites, into the network. At the present time, ECN is composed of 10 terrestrial sites with a further 2 due to join during 1994, and 37 freshwater sampling sites (fig. 2.3.4.1). In practice, the choice of sites has necessarily been governed by the more pragmatic of the above criteria. However, an analysis of the characteristics of the terrestrial sites in relation to the broad patterns of UK environmental variation reveals a reasonably good representation. The analysis used the ITE Land Classification system – a multivariate classification of all 240000 1km squares in Great Britain into one of 32 Land Classes based on a range of environmental variables (Bunce et al. 1981,1994). This provides an objective sampling framework within which to represent existing sites and to target locations for the recruitment of new sites. ECN plans to build a „spine" of representative sites, and then to recruit "associate" sites which would provide replication within the major axes of environmental variation.

The choice of ECN 'Core Measurements' (Sykes, 1994) was based on a selection of 'driving' variables and ecosystem 'response' variables, made by a working group of experts in each field. These variables are regarded as those sufficient for the operation of primary sites in the network; a list of 'priority additional measurements' has also been identified which are to be recorded if funds allow. A Statistics and Data Handling Working Group was convened from the outset of the programme, advising on site selection, sampling design, scales, resolutions and units of measurement. The Working Group, in association with a full-time ECN Statistician, continues to advise on spatial and temporal analytical techniques, for example mechanisms for interpolation of site-based data to a 'surface' of information, and methods of extracting patterns of environmental change from time series data.

ENVIRONMENTAL CHANGE NETWORK
FRESHWATER & TERRESTRIAL SITES

Fig. 2.3.4.1:
Environmental Change Network - Freshwater and Terrestrial Sites

Data are collected at the highest practical temporal, spatial and measurement scale resolutions to allow maximum flexibility in aggregation for comparative or summary purposes. Where national sectoral monitoring schemes were already operating (e.g., the Butterfly Monitoring Scheme, the Precipitation Composition Network, the Common Bird Census), their sampling design and methodology was adopted for ECN, with the advantages of analysing ECN sites with respect to broader regional and national patterns and providing a sample of sites having a comprehensive set of environmental information.

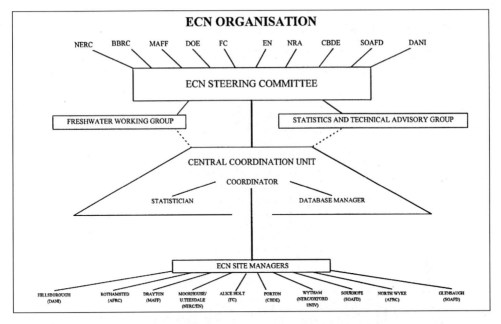

Fig. 2.3.4.2:
Environmental Change Network (ECN) Organisation

Where no such schemes existed, advice on methods was sought from scientists working in these sectors through the ECN Working Groups. Appropriate sampling design depends on knowledge of the variability of the parameter being measured – for example the number of replicates considered sufficient for water quality sampling from rivers. However, such information was not available for some ECN measurements. Ideally, pilot studies would have been carried out where necessary; however, lack of resources and a requirement to start the programme quickly meant that this phase was not possible. Nevertheless, ECN was able to use the first year of monitoring – 1993 – as something of a trial period and, where necessary, standard protocols have been altered for 1994 in the light of new information. In most cases, this has meant gathering more information rather than making significant changes. The present ECN Statistics and Technical Working Group, in association with the ECN Statistician, ensure that current and changing methods of measurement are themselves closely monitored, and that close contact is maintained with the scientific research community. ECN is conscious that monitoring gained something of a bad name in the past due mainly to a lack of such contact and of continual review.

ECN is managed by ITE, a component institute of the Natural Environment Research Council, on behalf of a consortium of contributing organisations, each of which is responsible for meeting the costs of the monitoring programme at one or more sites. The Central Coordination Unit (CCU), based at the ITE's Merlewood Research Station, consists of a Coordinator, Statistician and Database Manager. Figure 2.3.4.2 outlines current ECN organisation.

ECN Data Management and Quality Assurance

A section on the ECN database and data handling procedures is provided in a Workshop background paper (Sykes, 1994).

Any monitoring programme must be concerned with the quality of the data it generates. The most useful definition of quality is 'fitness for purpose'. ECN has adopted the following terminology in common use:

Quality Assurance: A term embracing all planned and systematic activities concerned with the attainment of quality.

Quality Objectives: The specification of target values for selected data quality criteria

Quality Control: The practical means of securing 'product' quality through actions taken at the data capture stage. These actions are part of the Standard Operating Procedures incorporated within the Core Measurement protocols, and form an important part of training programmes. Inspections should be held at defined intervals to ensure that procedures are being followed correctly.

Quality Assessment: Also called 'Quality Verification'. Procedures for assessment of the degree to which quality objectives have been met – ideally evidence that they have been met – after data capture. This needs to be a system of monitoring which feeds back into quality control. Data validation is part of quality assessment.

Objectives need to be set in relation to the long-term requirements of the programme. Quality criteria might include: accuracy, precision, resolution, completeness, comparability, representativeness, lineage. Where spatial data are concerned, an additional specification is useful to describe the structural requirements of the data; this is sometimes called 'logical consistency'. It relates to the topological relationships embedded within the data structure which may vary from a simple graphic 'spaghetti' (sets of coordinates with 'join' instructions) to one which incorporates information about associations between spatial objects (e.g. line definition X separates polygon A from polygon B).

Quality assessment procedures will vary according to the type of measurement. Some measurements can be kept or re-visited (e.g. invertebrate specimens, vegetation quadrats); these can be re-recorded independently and errors estimated. Other measurements are more ephemeral and cannot be revisited – for example meteorological data recorded at hourly intervals. Instrumentation errors may be capable of being back-corrected, but surveyor recording errors are less correctable, making quality control and training all the more important.

Data validation procedures screen data for possible errors which may have occurred at any stage during data capture and handling. These normally include range checks for numeric and categorical checks for non-numeric data. However, such univariate procedures may miss more subtle problems which nevertheless generate erroneous data within the valid range. Multivariate associations between variables and analyses of spatial and temporal series will pick up unexpected relationships which may be due to error (Scott, 1994). At present, all ECN data are subject to the first type of validation and problems are referred back to the sites for confirmation before any data are incorporated into the database. The second type of analytical validation is currently being implemented as a second stage process. Unless there is a very clear understanding of the error revealed or the unreliability of the data (e.g. known instrument malfunction), then the data are stored, but 'flagged' in the meta-database. Otherwise there may be a real possibility of throwing out values which are genuine outliers and indicators of change. Where data quality is affected by a factor beyond the control of the recorder, this information should also be stored. ECN uses a set of defined 'quality codes' which recorders can attach to the data records as required. Examples are: beetle pitfall trap flooded, trampling of vegetation plot, wet bulb wick frozen. This list of codes grows as common problems become evident. In addition, recorders may send free-text information about data quality; it is important that there are no barriers to such information flow.

The ECN database is managed by the CCU and is held at ITE Merlewood. At present it is a centralised facility, directly accessible over the network by authorised users. However, infrastructure and software are in place which will allow it to expand into a more distributed system where required. Integrated monitoring programmes generate not only large quantities of data but also complex databases having both spatial and temporal dimensions. The meta-information system forms an important part of the ECN database and includes dataset descriptions, quality specifications, units of measurement, spatial and temporal dimensions, details of associated material (for example: papers, maps, satellite images), and various reference information. Spatially referenced data are managed with a Geographical Information System (Arc/Info) which provides an internal link with the main relational database system (Oracle). Both software products are well-known and reliable and together provide many of the facilities for managing space-time data. However, programmes like ECN really require a much more comprehensive and integrated system of database and GIS facilities, with a range of programming and interface-building tools and built-in facilities for the management and analysis of time-series digital maps. Such systems are in development, but none has yet emerged as sufficiently stable and reliable for significant investment.

ECN is keen to develop links with other national and global monitoring programmes, and exchange information and experiences. The development of network database and meta-information systems will be one particularly important issue if national monitoring programmes are to form the basis for a global monitoring system.

References

BUNCE, R. G. H., BARR, C. J. and WHITTAKER, H. (1981) „An integrated system of land classification". Annual report of the Institute of Terrestrial Ecology for 1980, pp. 28–33
BUNCE, R. G. H., BARR, C. J., CLARKE, R.T., HOWARD, D.C. and LANE, M. (1994) „The ITE Merlewood Land Classification - the complete coverage of Great Britain" and „Land Classification for Strategic Ecological Survey". forthcoming publications
LANE, M. (1992) "Quality"Assurance within the ECN Programme". ECN working paper
SCOTT, W. A. (1994) "ECN Data Validation". ECN working paper (in preparation)
SYKES, J. M. (1994) "The United Kingdom Environmental Change Network" Background paper for UNEP-HEM Workshop on Harmonization of Environmental Monitoring (Annex II-3.10 this volume)

2.3.5 The Swiss concept of integrated ecosystem-related environmental observation: the ecosonde

K. Peter and P. Knoepfel

The background

In the late 1980's, complex environmental problems such as forest damage and the accumulation of pollutants in the soil led to the recognition of the urgent need for an environmental monitoring system which goes beyond sectorial boundaries. A better understanding of environmental processes and appropriate measures to overcome environmental problems can only be reached by way of a process which takes into account the cycles and chain reactions set in motion by various media. Thus, integrated environmental observation calls for the ecosystem perspective.

Since the coming into force of the Federal Act on the Environment (1985), the Swiss Federation and Cantons are under a legal obligation to provide a wide range of services in the area of environmental protection. However, experience has shown that the only measures which can be successfully implemented are those which are widely accepted by the population and local authorities, and only those solutions which are tailored to local circumstances will receive support at this level. This applies for both water and environmental quality measures and for the increasingly important areas of nature and landscape conservation. Due to their general binding force, national measurement plans tend to be greeted with opposition. The political solution sought to this problem takes the form of differentiated execution which takes account of regional political, economic and cultural characteristics. This kind of approach necessitates the differentiated evaluation and control of environmental quality. Thus, observation systems are required which take into account both specific local features, at the same time, however, there is a demand for activity at a representative nation-wide level.

Most of the national monitoring networks implement environmental monitoring on a sectorial basis. However, it is possible for different environmental sectors to be simultaneously affected by changes in the environment induced by measures introduced for one sector. The only way in which effects which occur outside the controlled parameters can be evaluated is through observation of the ecosystems.

The successful introduction of remedial measures is dependent on an awareness of the problem within society, and this in turn is linked with the perception of the environmental situation. For this reason, integrated environmental observation takes economic and social processes into account. As environmental problems do not stop at national borders, the national observation system should also establish links with observation systems in other countries. Ecosystem observation is particularly suited to this purpose.

The Swiss Commission for Environmental Observation

The Swiss Commission for Environmental Observation (SCEO) is a member of the Swiss Academy of the Natural Sciences. It was founded in 1988 with the task of developing a long-term formal system for environmental observation. As a part of this task it is also involved in the documentation of unexpected long-term changes in the environment and provides support for research into the causes of environmental problems and the interpretation of the necessary remedial measures. The twelve members of the commission were appointed by the Academy and selected on the basis of a broad, interdisciplinary knowledge base. Universities, research institutions, federal ministries and regional executive authorities are all represented in the Commission.

As part of its tasks, SCEO has developed a concept for integrated ecosystem-related environmental observation under the presidency of Peter Knoepfel and with the collaboration of four doctoral students. The proposal for this observation system, which has been available since 1993, meets national and regional requirements and was developed in tandem with comparable efforts in neighbouring countries [1]. The observation concept is intended to complement existing monitoring networks and surveys for the integrated and regional tasks that constitute environmental and nature conservation policy.

Integrated ecosystem-related environmental observation

The observation approach is based on the interaction at ecosystem level between the environment, the human race and its activities. Observation focuses on all relevant areas of the ecosystem, including socio-economic and political issues, and it is implemented in typical or exemplary areas (ecosystems). For practical reasons, observation activities must be restricted areas to exemplary areas. All of the major Swiss ecosystem types, for example, high mountain areas, small towns, rural settlements, will, however, be included in the selection. The measurement design realised as a result of the implementation of the observation programme in a specific observation area is known as the ecosonde. The ecosonde is the highly sensitive, area-specific measurement instrument to be used in integrated environmental observation.

The core of the long-term observation programme will consists of four basic subject-specific programmes. To ensure that individual areas receive adequate treatment, a minimum catalogue of suitable data recording methods with the relevant parameters, evaluation standards and quality control of observations is specified for each programme. The basic programmes were developed in areas where existing, long-term observation activities have shown clear inadequacies and/or where it is appropriate to aim at approaches that look beyond the specific area in question. The basic programmes include the areas of landscape observation, the biological programme, monitoring of material influences and observation of administrative bodies.

The data framework for the ecosonde is provided by existing environmental information and surveys. The ecosondes are not intended to replace existing measurement networks and surveys but to provide a complement to them; the ecosondes will use the existing networks and surveys in their observation of ecosystems. The ecosondes' typical data sources include aerial photographs, archive material, legal and administrative documents.

Ecosystem research, which provides new insight into functional links, methods and standards for the routine operation of the ecosondes, goes hand in hand with the observation. The ecosonde provides ecosystem research with a well-equipped, interdisciplinary data framework for the generation of hypotheses and for testing.

Environmental information systems represent an important integrative tool for ecosystem-related environmental observation. Information to be stored in the database should include more than the actual measurement data: the documentation of target questions, hypotheses, methods, models and sources is also important.

Applications

Integrated ecosystem-related environmental observation with ecosondes provides the following services for implementation of measures and policy planning:

- It enables comprehensive evaluation and monitoring of the environment.
- It provides environmental research with a well-equipped long-term data laboratory.
- It enables the timely recognition of damage (early-warning) and new insight into connections between the causes of damage as well as the comprehensive evaluation of effects necessary for integrated environmental reporting.
- In addition to a control of success of political measures, the results also provide a basis for the development of interdisciplinary policy programmes and for the evaluation of effective measures.
- The ecosonde provides a management instrument for the implementation of regional executive tasks.
- The ecosystem-related approach makes it possible to integrate global issues with regional requirements.

Steps towards the implementation of the ecosonde

Ecosonde observation exists as a concept. Prior to its general implementation, it is intended to verify the feasibility of the proposal on the basis of a concrete example.

Following successful examination of the concept and verification of its use in the fulfilment of regional, national and scientific requirements, an application will be made for its adoption as a permanent regionally based monitoring system as a complement to existing observation activities at national level. The Environmental Observation Steering and Coordination Organ *(Lenkungs- und Koordinationsorgan der Umweltbeobachtung – LEKUB)* established by the Swiss government in 1993 will be given the task of examining the application for permanent institutionalisation of environmental observation systems in Switzerland and by doing also make proposals concerning the ecosondes.

It is estimated on the basis of the variety of natural landscapes and ecosystems in Switzerland, that 8 to 12 ecosondes will ultimately be required.

The commission is at present involved in the preparation of the planned feasibility study which is intended to start in 1995 and continue for a good three years. It is intended to complete the study by spring 1998. Following successful completion of the project, permanent operation for the ecosonde observation can be planned for the test area used in the feasibility study.

The feasibility study

The aim of the planned study is to examine the feasibility of the ecosonde concept on the basis of a concrete example. The tasks to be addressed include:

- The implementation in the test site of the four basic programmes on a pilot basis using the principle methods suggested in the programmes
- The integration of the basic programmes with existing observation activities and data surveys. The processing of previous surveys and information for use in integrated evaluation (harmonisation of data: balancing with help of statistical models in the arc/info.).
- The compilation of a database with the possibility of integrated and ecosystem-related evaluation: access to the database will be available to several users.
- The identification of key processes (core cycles and chains of effect) for the area observed.
- The evaluation of models, calculation and statistical methods for the description of the links relevant to the execution of measures.
- The processing of results in relation to environmental reporting which presents links between effects and complex relations in a comprehensive form.
- The establishment of an ecosonde measurement design for the test site to be used as a basis for permanent monitoring.
- The adaptation of the selection of measurements to international observation programmes (Global Terrestrial Observation System, observation of biosphere reservate) and for comparison with ecosystem-related environmental observation in neighbouring countries (e.g. Germany).

The feasibility study is an interdisciplinary project which will combine operative observation activities with research. It is also rooted in a varied environment ranging from international links to the regional base. The project planning already contains provisions for the organisational needs arising from this broad background and scope. The project will be financed by federal ministries, Swiss National Scientific Fundation, Cantonal authorities, private sources, which are also future users of the observation system. The strategic management is in the hands of a management committee which will include representatives of the most important clients. The operational project management will take responsibility for the coordination between the clients (management committee) and the project group and also the operational monitoring of deadlines, costs and targets. The data processing department (database) will take responsibility for integration in the specialised areas. The subject areas will be developed by the institutions (technical research centres and universities) and private concerns with the joint target of integration.

Regular reports on the status of the work in progress shall be submitted to the client at the end of each processing phase. Operational target control shall be the task of the project management. The monitoring of the project progress and strategic decisions will be the responsibility of the management committee whose members include representatives of the sponsors.

The time schedule is based on the research application to be submitted to the Swiss National Research Fund for the financing of the research component. The project is divided into three phases. The preparatory phase, during which existing data will be merged and the database and integration developed, will start in 1995 and will take six months. Approximately two years is allocated for the following processing phase which will involve the execution of the basic programmes and the integrated evaluation. The implementation phase, during which the results of the integrated evaluation will be studied from the perspective of their use in implementation, will take a further six months. The cost for the entire project for the duration of three years is estimated at a total of 2.5 million sFr.

Focal points of the feasibility study

The Swiss Commission for Environmental Observation is convinced that active interest and support are important prerequisites for the execution of a feasibility study. Representatives of the Aargau cantonal authorities have supported the Commission's work for some time now as they are interested in filling the gaps in current observation activities and in demonstrating the solution of urgent coordination and integration of environmental information between and within national and cantonal measures. There are several alluvial sites of national importance in the canton of Aargau. The department for nature and landscape protection of the Federal Office of the Environment is planning a programme to monitor the success for the alluvial site inventory nation-wide. The Alluvial Site Decree recently came into force. The protection and evaluation of alluvial sites are also high-priority concerns on the international stage.

The observation area has already been defined in cooperation with the canton of Aargau and the Federal Alluvial Site Consultancy. The area in question is Umiker Schachen near Brugg which includes an alluvial site of national import. A number of surveys and studies already exist on Umiker Schachen which could be integrated into the evaluation as part of the study. Valuable synergy effects could be released through the integration of the two observation projects (the feasibility study ecosonde and the monitoring of the alluvial sites). The inventoried alluvial sites represent 0.25 % of the total area of Switzerland and provide a habitat for 40 % of Swiss flora. The inclusion of alluvial sites in the feasibility study means that it will be possible to check whether integrated, ecosystem-related environmental observation is suitable as an instrument for the monitoring of biodiversity.

Note

[1] GROLIMUND, P. und PETER, K., 1994: Integrierte ökosystemebezogene Umweltbeobachtung, Konzept für die Einführung eines Beobachtungssystems, vdf Zürich, in preparation.
Short report of 3rd March 1993 by the Swiss Academy of the Natural Sciences (in German and French)

2.3.6 Some principles for the organisation of GTOS

J. Nauber

For many reasons the international community has decided to elaborate Global Environmental Observing Systems. This is a reaction on the growing concern about the state of the environment and the realisation that too little is known in order to scientifically judge its quality and to make recommendations on environmentally sound behaviour of the international community, national governments and individuals. The question to be answered is which behaviour causes which effects (changes) on the different scales of the environment and vice versa how do these changes impact human life. The answer to the question should give mankind the basis for finding out measures in order to improve the current situation.

A part of a global observation system is the establishment of a system for monitoring and interpreting changes in terrestrial systems which is complementing the already existing systems for ocean observations (GOOS) and climate observations (GCOS). Harmonization and coordination are an absolute prerequisite for this system as monitoring activities are widely distributed concerning space, time and thematic orientation. Other reasons for coordination are

- Global budget constraints ask urgently for most cost-effective systems; coordination must be seen as the realisation of the principle of austerity. The German government consequently has considerably contributed to the establishment of HEM (since 1989) which might be seen as a possible basis for a coordination instrument for the future. This question has not yet been decided and will be part of the ongoing negotiations between UNEP and the German Government once UNEP has presented its overall strategy for EARTHWATCH.

- Users need one partner entity which supplies the desired results in an adequate aggregation level and which includes the requirements of the user side and feeds them it into the actual monitoring activities.

- The latest development of data management indicates that the times of huge centralized data bases has passed as electronic communication makes data available nearly at any place of the world at any time. It must be the principle that on the one hand monitoring activities are executed as decentralized as possible. This requires on the other hand an agreement on harmonized measurement techniques and the functioning of a meta database which gives information about who is doing what, the results of projects and how to get involved in certain activities.

- The organisation of GTOS must be a combination of top-down and bottom-up approach. Top-down indicates the encompassing fields of interests (global and regional approach) while bottom-up is directed from local to regional issues in order to enter those problems into GTOS which have direct impact on local changes. Additionally the local level gives indications about the potential of sites for participating in GTOS, as is the case for example with the Biosphere Reserves Integrated Monitoring Programme (BRIM) (see Annex II-3.1).

My conclusion from what is mentioned above is that a coordination/harmonization organisation should be foreseen in the structure of GTOS and the decision soon has to be taken whether HEM could be an adequate institution to be entrusted with this task which of course has to be defined in detail later on. That is a question which depends on the not yet finalized deliberations of UNEP. This issue will be subject to further discussions between those involved.

2.3.7 Challenges for developing countries

L. O. Ogallo

Introduction

The environmental processes controlling the space-time characteristics of the terrestrial ecosystems are often quite complex, and highly interactive with the general processes of the atmosphere and the large water bodies within the neighbourhood. Some of these interactions involve complex natural biogeochemical cycles, many of which are not well understood.

Changes in the space-time patterns of terrestrial ecosystems have been the subject of many national regional and international monitoring and research programmes due to the close associations between the ecosystems and many socio-economic activities. Such changes have been associated with negative anthropogenic and natural forces. The negative anthropogenic forces have included the overutilization of resources, degradation and mismanagement of the traditional environment which supports the natural cycles of the various ecosystems, among others. The negative natural forces on terrestrial ecosystems include extreme changes in the traditional conditions of the local environment like those that are linked to prolonged extremely wet/dry, hot/cold climate change, and other hazardous environmental conditions.

For sustainable use and management of the terrestrial ecosystems it is therefore crucial to monitor on a long-term basis the space-time changes in the patterns of the ecosystems, together with the associated environmental and anthropogenic processes. The proposed Global Terrestrial Observation system (GTOS) is based on this principle.

GTOS is expected to a large extent to utilize the available observational systems. Most of the existing observations are however driven by sectoral interests, resulting in some differences in the sensors used, methods of data collection, processing, validation, dissemination and archiving. The success of GTOS will depend on its ability to harmonize such multisectoral and interdisciplinary data sets for a wide range of users with different interests, requirements and objectives. The following sections will highlight some of these challenges.

Vision

The overall vision of GTOS will be the major guiding factor to the proposed data harmonization programmes since it will specify the major priorities of GTOS at short medium and long-term scales. Some of the various sectors with observational platforms have long-term programmes spanning sometimes into the next century. GTOS interests could be included in some of the future programmes with minimum cost, especially with the current level of high technology.

User Requirements

The basic objective of GTOS is to provide timely, accurate and credible environmental information to a broad user community including the decision- and policy-makers. This information is often more crucial at national levels where effective management of ecosystems and environment are required. The international, regional and national requirements in the harmonization processes must therefore be clearly be identified.

Challenges for the Developing Countries

Data harmonization normally involves the comparison of a number of data acquisition stages which include method of measurement, data processing, quality control, dissemination archiving, etc.

These require good, sustainable software and hardware, telecommunication facilities, a strong national/regional base (focal point), human resources, training and education on the basic harmonization methods, running cost, etc. Most of these are lacking in many developing countries, especially large parts of Africa. The concept of total use of the products derived from global/regional centres at national levels must be discouraged due to the uniqueness of ecosystems and environment at local levels. Institutional framework must therefore be developed for national, regional and international networking of GTOS observations.

Conclusion

The data harmonization programmes must be integrated with the overall short, medium and long term virsions of GTOS. The harmonization programmes must also take into consideration the unique regional and national requirements which serve a wide range of user communities with different interests, requirements and objectives.

For the harmonization programme to effectively contribute to the GTOS mandates, special consideration will have to be given to some of the developing countries, especially in Africa in terms of software, hardware, telecommunication facilities, human resource development among many others. An institutional framework for networking the national, regional and international GTOS nodes must be part of the overall planning of GTOS.

2.3.8 The Russian view on forming GTOS

Y. A. Pykh

Introduction

From my point of view, one of the most important questions is the compilation of a Russian sources Data Base and its integration into the international environmental community.

The first step in this direction is the formatting of a data-base of Russian Research Institutes, organizations, national programs and projects which deal with different types of terrestrial information.

The Center for International Environmental Cooperation (INENCO) is in a good position to format such a data-base and to integrate it into the international environmental community of information.

Another important question is the elaboration of an internationally adopted, computer-based procedure to harmonize environmental data for use in various kinds of terrestrial computer models, formatting of sets of environmental indicators and the organization of decision-support systems.

Use of ecosystems modelling for GTOS

One of the most powerful tools that could make it possible to recognize cross-sectional linkages of environmental processes, the overlap in interest areas of different environmental programmes and activities, and interconnection between activities at different levels and different fields is the various kinds of mathematical models.

INENCO Center has a long and successful experience in elaborating mathematical models of various kinds of ecosystem processes and ecosystem dynamics in general.

Let us give a very brief description of the main trends of ecological modelling. First of all there are models of pollutant dynamics in the ecosystems, including radioactive, organic and non-organic contaminants. These models with the general title POLMOD describe the contaminant dynamics in the ecosystems including lower atmosphere, soil, vegetation, surface and underground water and hydrobionts. The processes of accumulation and destruction of contaminants are analyzed on the basis of existing concepts and ideas of chemical migration in the ecosystems. The method of response functions was used in the description of the processes. In each case, parameters of the models were evaluated and tested on a wide range of literature and experimental data for the ecosystems of various geographical zones – from middle taiga to deserts.

Another type of model is that of such ecosystem processes as ontogeny in higher plants (agricultural) and humus dynamics in natural and land-used ecosystems. The model of plant ontogeny allows the prediction of rates of growth and development processes. Only the most favorable combination of these two main physiological processes results in the optimal yield. The main input parameters for this model are typical weather and soil conditions for the given site and type.

The model of humus dynamics was done for various soil types in natural ecosystems as well as under the impacts of various types of agricultural land-use and global climate change. The driving parameters of humus formation in the natural environment are analyzed and described in this model. The special erosion submodel calculates the additional loss of humus due to water and wind erosion. The main practices of agricultural land use resulting in a change of humus dynamics and the scenario for global climate change impact are demonstrated as simulation experiments. The evaluation of parameters and testing of the model are given for 25 soil types from tundra to desert ecosystems.

Harmonization in the context of environmental monitoring seeks to bring together various kinds, levels, and sources of data in such a way as to be comparable and compatible and thus useful in decision making. Harmonization is involved throughout the processes from observation to policy setting, i.e. from data collection, to the development of information, data integration and aggregation, as well as formation of environmental knowledge for a common basis of decision-making. Effective harmonization requires a good knowledge of the nature of the phenomenon being measured as well as a clear concept of what goal is to be achieved.

The completion of the general processes of data harmonization, from our point of view, and the fulfillment of the main part of the above requirements could be done on the basis of the theory of Environmental Indicators. Such an approach is also being elaborated on now in our Department of Mathematical Modelling. Based on the more complicated models described above, this approach demonstrates the results of ecological forecasting and data harmonization in the most convenient way for decision-makers.

Geographical information systems for GTOS

Geographical information systems (GIS) have become very important in the storing, evaluating, depicting, updating and processing of spatial data.

Taking this into account I suggest the following GIS- based items for the Russian Federation:

1. Analytical compilation, running and regular updating of the Register of Russian sources/data bases relevant to GTOS.
2. Analytical compilation, running and regular updating of Russian research and monitoring programmes relevant to GTOS on a Federal level and at the level of Agencies – Academy of Sciences, Ministry of Environment, Ministry of Defense, Ministry of Oil and Gas, other industrial Agencies, Ministry of Agriculture, Agency for Hydrometeorology and Monitoring of the Environment, Agency for Water Management, Agency for Forestry, Agency for Fishery, Ministry of Geology and others, and on a level of (selected) regional administrations and urban megapolices.
3. Consultation and expertise on actual data availability, actual data quality and on current mechanisms of access to data bases widely dispersed among numerous Russian agencies and institutions, including technological aspects of data flow.

4. Current dissemination of information and materials relevant to GTOS activities among Russian institutions and agencies, including public relations issues.

Global climatic change impacts on the environment

Global climate warming is one of the most important factors of present global change. The changes in global and regional thermal regime and atmospheric precipitation resulting from carbon dioxide and other greenhouse gases have strongly influenced the environment, its stability, ecological security as well as the sustainable development of society.

A research group from INENCO Department of Global Climatic Change Impacts on Environment is dealing with the specific issues of the modelling of terrestrial ecosystem dynamics resulting from various types of anthropogenic impacts. Primarily, these are connected with the changes of atmospheric gaseous composition, air pollution, climates and weather changes as well as human economic activity.

The current activity in this area is as follows:

1. The modelling of terrestrial ecosystem dynamics and the evaluation of the role of different factors in their development and degradation;
2. The estimation of anthropogenic carbon dioxide and atmospheric trace gases impacts on photosynthesis, vital activity and productivity of natural ecosystems;
3. The study of the influence of global warming and regional climate changes on agriculture;
4. The study of the influence of global climate changes in natural zonality;
5. The study of polar and mountain permafrost degradation under global warming;
6. The modelling of ecosystems sinks and source changes of carbon dioxide and methane under global warming;
7. The study of ecological and social aspects of land-use and land-cover problems.

INENCO's Department of Global Climatic Changes Impacts carries out wide research in the areas of ecological modelling and the impact of climate changes on agriculture, which are included in the following national programmes:

a) Ecological Security of Russia;
b) Global Climate Changes and Their Consequences;
c) Paleoclimatic Scenario Using the Future Climate Changes;
d) National Report for the US Frame Climate Convention and others.

The Department has a permanent collaboration with many international and bilateral working groups and projects. This includes:

- Inter-governmental Panel on Climate Change (IPCC). Agriculture and Forestry;
- YIII Working Group of Russian-American Agreement on Environment;
- Russian-German Project on ecological Modelling for Monitoring of Forest Reserves;
- Russian-Argentinean Project on Climate Change and their Consequences;
- Russian-Japan Project in Carbon Dioxide Sources and Sinks, and others.

Proposal

The INENCO Center could carry out investigations for GTOS on the problem of harmonizing ecological models. This will encompass the main aspects of the problem, starting with the harmonization of model input parameter sets and the inter-comparison of research procedures and algorithms, finishing with the output formats and intermodel harmonization.

2.3.9   Possible contribution of the Nature Reserves Authority (Israel)

J. Cohen

Data sharing

Since the Global Terrestrial Observing System (GTOS) will try to rely on existing data sources, the availability of data and the possibility of sharing it, is of crucial importance. Data "owners" should be encouraged to make their data available. The use of aggregated data for GTOS will enable access to sensitive data by passing such confidentiality problems related to ownership / economic issues.

Monitoring sites and regional/country data centers

It is suggested in an early stage of GTOS, that in each potential participating country, an organization that might have long-term commitment to GTOS should be found. Then, in coordination with the central effort of GTOS, it is possible that the national organization will be able to help identify the available data sources, the possible monitoring sites and the available monitoring facilities. The steps needed to be taken in order to obtain this information, could then be recommended to the other GTOS participating countries. In a later stage, the national organization could lead the monitoring program in its own country and the connections with GTOS and the other GTOS partners.

It is suggested that these organizations will be approached by UNEP through their foreign ministries, thus giving some incentives to the governments to support GTOS activities on a long term basis. In some cases, external support will be needed, both in establishing and running GTOS activities.

Some of the most important monitoring sites need to be located in zones of Ecological transition, where changes might be more significant.

Since Israel is located in such a transition zone between the Mediterranean forest on one hand and deserts on the other, it is possible to locate two monitoring sites, that at a distance of 150 Km will represent different ecosystems. One site could be in the proposed Biosphere Reserve Mt. Carmel (Mediterranean forest) and the other could be located in the Negev desert, close to one of the Nature Reserves.

General responsibility and operation could be given to the Nature Reserves Authority of Israel (NRA), which is responsible for all the Nature Reserves and nature conservation in Israel.

With some international and national support, I believe that the NRA could organize to serve as a center for GTOS activities, and to coordinate the required activities, scientific or otherwise.

2.4 Constraints to the extraction of information from complex environmental datasets

I. K. Crain

Introduction – The complex environmental database

The environment surrounds us as we move through life; it affects our personal health and prosperity, as well as the well-being of regions, nations and the globe. "Environmental Information" or "Resource Information" therefore has an equally encompassing scope. Of primary importance to decision-making is quantitative information, derived and integrated from measurement and observation which can tell the decision-maker how much, how many and how fast changes are occurring, and whether established standards are being met. The subject-matter scope extends well beyond the obvious of ecology. The World Commission on Environment and Development noted "The environment does not exist as a sphere separate from human actions, ambitions and needs...the environment is where we all live; and development is what we all do in attempting to improve our lot within that abode. The two are inseparable." (WCED, 1988) Through this inseparability comes a total information package of environmental information which must include development and human information – that is economic. and social factors.

The emerging paradigm for sustainable development of the Earth's resources was sketched out by the Bruntland Commission in 1987 and further elaborated recently in Agenda 21. There is the frequent occurrence of such indicative words and phrases as "global", "integrated", "eco-systems", "human factors", "equity", etc. It is no longer possible to conduct resource assessment on the strictly sectoral basis of resource economics. The key is an ecosystems approach, in which human activity and well-being are an integral part. Mathews and Tunstall (1991) of WRI have used the term "eco-development" for this paradigm. This paper summarized well the changing perspectives over the last few decades from "Frontier Economics" (the myth of the infinite resource) which governed developed countries until the 1960's, through the "Environmental Protection" era of the 60's and the "Resource Management" paradigm of the 70's and 80's which took a sector-by-sector approach to resource sustainability, to the new era of "Eco-development". In their words, *"Born of both resource management and the principles of ecology, this approach reflects the beliefs that economic systems must harmonize with ecological systems and that synergy must reign between environmental management and economic development"*. This paradigm embodies the idea of sustainable development, but from an eco-systems perspective, rather than simply resource economics. This approach requires that the base of data for sustainable resource decision making has the characteristics of being **multi-sectoral** (including both environmental and socio-economic data), **spatial** (often over very large areas of the globe), and **diversified** (containing statistical data, quantitative scientific measurement, nominal and classified data, and descriptive/narrative information).

A list of contents of the information package for the environmental decision-maker must include:

- **ecology** (encompassing climate, wildlife, vegetation, soil)

- **economic activity** (inputs, products, wastes, wealth and resource stores)
- **social information** (including population and health)
- **abiotic condition** (including the shape of the land)

Given this vast scope, an inconceivable volume of data exists. Much of it is collected locally for operational purposes far removed from policy decision-making. Recent advances in satellite remote sensing enable the daily capture of vast amounts of new data on the atmosphere, biosphere and lithosphere – far more than will ever be analyzed or refined into "information"

Even if one considers a narrower sector such as forestry, monitoring and assessment in an eco-development paradigm cannot ignore the wildlife, the water, the soil, and the human populations which exist in and near the forested areas. The scientific assessment of such massive data assemblies (even within a specified sector) and the subsequent decision making process must have the technological tools to integrate, abstract and summarise the information

Information extraction – The challenge of decision-maker needs

Policy decision making involves three stages; information gathering (called "Intelligence" by Newall et al, 1958), identifying alternative courses of action and their consequences (the "Design" stage), and finally the Decision, based on optimizing the consequences (often in consideration of subjectivity, such as "social well-being"). An important characteristic of the Design stage is that it depends heavily on data banks and the models that operate on those data banks to forecast outcomes of proposed decisions.

The nature of the policy decision-making process and the enormous scope of environmental information imply the need to reduce these data banks (or databases) to a manageable size through both summarization and abstraction to a relatively high level (Fig. 2.4.1).

Summarization reduces the information overload by grouping large numbers of observations and representing them by a single quantitative value. Time-series may be summarised by annual averages e.g. mean annual temperature, and spatially distributed data may be averaged over large regions. One common framework for the latter is by administrative region. For environmental information, however, there are disadvantages to this approach. Administrative boundaries seldom align with the natural boundaries which encompass ecological zones. National or regional summarization of statistics can dilute or mask the magnitude of change.

The ideal framework for summarization from an environmental viewpoint would use natural boundaries defined for instance by watersheds, or eco-climatic zones. An important corollary of this approach is the need for information management tools which can integrate, correlate and transfer information to/from the administrative frameworks familiar to decision-makers from/to more issue oriented frameworks.

Abstraction reduces the volume of information by selecting a few classes of observation as proxies for the behaviour of entire systems. These proxies or indicators can have many instances, i.e., be measured at many locations of the world at frequent intervals. The search for a single indicator which would reflect the health of the environment or an ecosystem in the way that the Gross National Product (GNP) is supposed to indicate the health of a national economy, is probably fruitless. Many environmental phenomena are spatially non homogeneous and so, the needed indicators must be multivariate and spatially distributed.

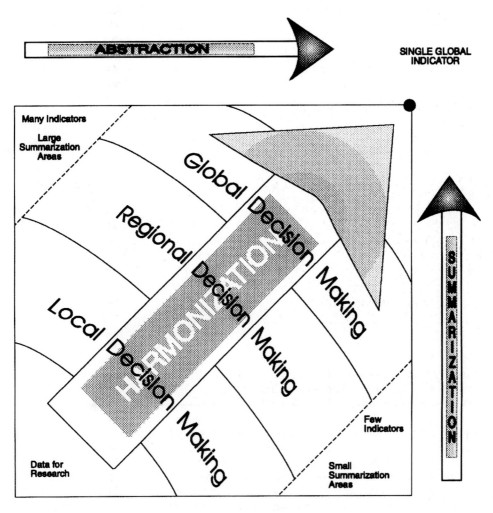

Fig. 2.4.1:
Abstraction, Summarization and Harmonization

The degree to which abstraction can be employed to reduce information to a few indicators is highly problem dependent. Complex issues need complex information and over-simplification does not help.

Large and complex databases are needed to provide a dear and reliable picture of the state of the resources and socio-economic conditions to facilitate decision-making processes at all levels. In order to obtain a comprehensive and reliable regional or global view it is necessary to bring together information from a broad range of resource assessment and monitoring programme. There are two major obstacles to achieving this . The first is the problem of data comparability and compatibility, including consistent approaches to data aggregation, to ensure that data remains comparable at different level of summarization. The second is that a comprehensive view a the environment requires an integrative modelling approach and a comprehensive set of measurements covering different media, different biomes, different ecosystems, different regions, etc. – and at present there is little overall coordination between programmes or an agreed framework for the rational integration of diverse resource databases. Ensuring that data collected by different institutions, under different programmes, and for different purposes is comparable and compatible, is the task of "harmonization" (Keune and Crain, 1993).

Some progress is being made in the development of the necessary ecosystem framework to form a basis for applying the technology of large resource databases. Since integrated approaches to environmental assessment have been developed h a number of programmes a local scale in the fields of ecosystem research and monitoring, these are likely to provide the best starting point for development of an integrated global framework for environmental research and monitoring. A similar approach is being considered on a regional scale by the EEA Task Force (EEA, 1992) which has recognized the need for an integrated framework for the provision of the environmental information needed for decision-making. Harmonization and standardazition of all activities, from measurement through management, are recognized as a central theme. Although details vary, the general approach to such integration is seen as involving such components as:

a) ecosystem research spatially oriented biological, physical, chemical, and geological research to understand the structure and function of ecosystems and to develop models of interrelationships,

b) environmental monitoring spatial monitoring of representative types of ecosystems and their components for ecosystem-oriented environmental assessment and management

c) environmental specimen banking: storage of representative environmental specimens for documentation and retrospective analysis, and modelling.

The experience and knowledge gained using such approaches can be used as a basis to develop an integrated global framework for environmental research and monitoring, although considerable work will be necessary in order to develop these ideas in detail.

A companion to this integrated framework is the need to find means a combining and "harmonizing" data of variable quality and uncertainty. Until standardization of monitoring is more prevalent, it will be necessary to "make do" with the best available, but recognizably incompatible and inconsistent data.

In summary, the primary challenges to the use of information by resource decision makers are as follows:

- Effective management of very large spatial databases
- Abstraction and summarization in a manner which facilitates decision making
- Integration of data from different environmental and socio-economic sectors
- Representation and management of error and uncertainty in the data

Constraints - the status of environmental information management technology

Database Management Technology

It is obvious that environmental information for policy decision-making will be voluminous, even given considerable summarization and abstraction. Data management technology has had a number of recent advances of potential beneficial application. Until the mid 1970's database management systems (DBMS) were commonly either of a hierarchical or network structure. These data models are restrictive and so database structures lacked the flexibility needed by decision-makers.

Recent years have seen the established operational use of "relational database management systems". RDBMS organize the information as a set of tables or "relations" linked by relational keys. Relationships between files are implicitly determined by common values in the "related" tables. Thus the RDBMS offers a flexible database structure which is easy to define, expand and change. Dynamic relations are created at query permitting considerable exploration of relationships, typical of the needs of environment informational analysis.

Associated with RDBMS, has been the creation of "Structured Query Language" (SQL). This is an English-like retrieval language which adopts a mathematical set theory concept and uses the same symbols in query definition. This allows for easy use of various data tables independent of the brand-name of the RDBMS, thus considerable improvement in the potential for information integration.

The RDBMS conception is not without limitations, however:

- it is not well suited to holding spatial data, such as coordinate strings
- its "table" based structure makes it difficult to deal with information which is not coded with a limited vocabulary
- it can be awkward to deal with hierarchical classification schemes

One new approach which seems more suited to the nature of environmental information is the Object-Oriented Database Management System (OO-DBMS) which could give improved capability in dealing with groups of data items (objects) and connecting between objects (e.g. land polygons) (Premeriane et al, 1990). The OO approach provides an intuitive and powerful way of creating and implementing data models of complex systems such as ecosystems. At the present time, this technology is immature and therefore suffers from the poor performance and lack of standards which plagued the RDBMS a decade ago. The policy decision maker must look to the future for the potential of this technology to allow for improved information compatibility on a broad base.

Another reality of complex environmental databases is that their components are likely to be held at physically distributed locations around the world and employ technology which, while optimum for the sectoral component will likely vary with the custodial institution. There have been recent developments, as well, in the so-called "distributed database systems". A distributed database management system (DDBMS) is a single logical database that is physically distributed on several computers, communicating through a network. It provides multiple access to the database and mechanisms to avoid conflict in update, retrieval and backup of data. DDBMS can be very useful for very large databases where most users are concerned with only a small subset of the database, while permitting access to the user with broader needs. In reality, existing databases are hard to link in a DDBMS because of differing technology at each site. A solution to this which is now developing is the "Federated" database management system. It shares the advantages of the DDBMS, while additionally allowing data communication among DBMSs of different data models, brand-names and access methods. Each local DBMS is autonomous in that it preserves its own characteristics as viewed by the local user. The complex environmental database clearly needs this technology, to link heterogeneous information collections in various countries and agencies. The technology is expected to mature in the next 5 to 10 years, but currently, linkages between heterogeneous databases must be made through agreed upon "standard" interchange formats, which continue to be elusive, especially for spatial data.

Geographic Information Systems Technology

Geographic Information Systems (GIS) technology has been used to manage environmental information and aid decision-makers since the inception of the Canada Geographic Information System (CGIS) in Canada in the mid 60s (Tomlinson, 1987). The basic advantage of a GIS is its ability to manage and perform complex processing on the spatial component of the data as well as the statistical aspects. In that way the actual geographic boundaries of regions defined on the earth's surface can be manipulated. This allows for the integration and summarization of environmental information using natural units – such as watersheds, natural forest areas, soil units and so on, and to combine these effectively with man-made administrative data collection units. Thus it can provide the link between the decision-maker's viewpoint and the natural boundaries of the problem.

In spite of the growing interest and demand, current GIS technology is limited in its applicability for environmental decision-making. First, regional as well as global spatial databases are frequently so large as to exceed the capacity of existing systems (Crain, 1990), and second, it is commonly necessary to store and process in excess of 100,000 spatial objects (e.g. soil units) requiring millions of coordinate pairs or raster cells to define. These two factors – sheer data volume, and the need to perform repeatedly complex geometrical computation on spatial objects – are the cause of the practical upper limit on the useable size of spatial databases in current GIS. Even with only moderately sized databases, users commonly note "lack of user friendliness", "slowness" and "unreliability" (Kalensky, 1988).

In addition, most GIS have little means of representing or processing information which is uncertain or of variable quality. All spatial and attribute information is taken as equally accurate, and subsequent processes do not propagate or analyze the effect of error or uncertainty.

A further restriction on usefulness is the lack of standardisation of internal structures and access methods. Current proposed standards for spatial data interchange contain little to help in communicating the data structure or the classification system(s) used for the data. No standard spatial access protocol like SQL is yet on the horizon, so it may be some time before the heterogeneous distributed GIS needed for effectively supporting decision making is a reality.

Data Integration and Presentation Tools

Effectively combining information from different environmental sectors requires knowledge of the measurement techniques, standards, classification systems, quality assurance methods, and terminology used in obtaining and describing the information. This associated data or "metadata" must be maintained and organized for use in integration and hamonization models on an ecosystem framework. Modem meta-databases are now using hypertext linkages and multilingual thesauri to form flexible relationships between the information descriptors (Benking and Kampffmeyer, 1992). This could assist in providing guidance to the decision maker on locating the appropriate information, on its quality and accuracy, and its appropriateness to models. Such meta-databases are the key to an effective intelligent human interface to an environmental decision support system. Although a few small steps have been taken in this direction, including those of HEM with HEMDisk, there is little standardisation or even agreement on conceptual requirements for metadata.

Particularly in the area of ecology, much of the data has a subjective element, and considerable uncertainty. This results from the intrinsic complexity and randomness of nature, combined with finite budgets (one cannot afford to measure everything, everywhere), and so can never be entirely removed. Decision making must always take place in the presence of imperfect data. The technology of uncertainty management and its presentation to decision makers is therefore essential. There have been some recent advances in this area, for instance, employing fuzzy set theory and fuzzy logic to manage retrieval in databases with subjective or "verbal" attributes (Lam and Crain, 1992), and the use of spatial uncertainty zones and error propagation techniques to present visualisation of relative uncertainty to decision makers in the form of an uncertainty map (Crain et al, 1993). These are however experimental results, and no useable system is currently available which can routinely present a "picture" of relative uncertainty to accompany the results of spatial modelling. This is in part due to the lack of a good theoretical base for the nature of uncertainty and its propagation in natural or ecological information.

The ability to effectively summarize and integrate environmental data is also hampered by the scarcity of appropriate models of ecological and natural resource processes, which would enable answers to questions of:

- what does this (information) mean?
- what if ... ?

Models are needed in such areas as climate change, ecosystems impact (of management decisions), biodiversity, human health and environmental relationships, error and uncertainty propagation etc. Where these models exist, information is needed on their scope, reliability, limitations, robustness.

The current methods of presentation of the results of models to decision makers tend to be unimaginative and inappropriate. Few methods seem to be available (or used) to effectively integrate the tabular view adopted in socio-economic studies with the cartographic view favoured in resource and environmental studies. One seldom sees the use of three-dimensional visualization, maps integrated with business graphics, interactive "what if?" systems etc.

Constraints – a summary

- difficulty in dealing with complex spatial objects in to-day's database structures (therefore must oversimplify)
- difficult to integrate data from differing database technologies
- cannot identify linkages in non-coded (e.g. narrative) information
- GIS limited in size of database that can be processed at one time

- difficult to integrate data from differing GIS structures
- the low-level "tool box" approach of GIS is not suitable for decision maker support
- cannot represent uncertainty or use "fuzzy" retrieval
- limited modelling capability of GIS and general lack of reliable models
- lack of effective methods of presentation to decision makers with integrated "views" and visualization

Progress – suggestions for working within the constraints

- ensure that terrestrial monitoring is systematic (and thus summarization is possible and meaningful)
- select and standardize key measurable parameters – reducing the dependence on classification systems (thus putting a scientific basis to the abstraction process)
- develop good meta–databases on available data, models and effective methods of data integration
- establish a systematic information management framework for terrestrial information
- exercise systematic quality management and provide quality information with data
- identify the information users and their needs and establish monitoring programs to address those needs
- cooperate

The future – suggestions for breaking the constraints

- encourage the development of spatial decision support software which incorporates multiple "decision–maker views" and use, experimentally, currently available systems of this nature
- develop, document and utilize good natural resource models and models which relate human factors to environmental change
- educate decision makers on the availability and use of existing tools
- pay as much attention (e.g. funding) to information management, including quality assurance, as to data collection

References

BENKING H and KAMPFFMEYER U, 1992, Access and Assimilation: Pivotal Environmental Information Challenges, Geojournal, vol 26, no. 3, pp. 323–334.
CRAIN IK, 1990 Extremely Large Spatial Information Systems –A Quantitative Perspective, Proc. 4th International Symposium on Spatial Data Handling, Zurich, July 23–27, IGU/IGA, pp. 632–641.
CRAIN IK, GONG P, CHAPMAN MA, LAM S, Alai J, and HOOGSTRAAT M, 1993, Implementation Considerations for Uncertainty Management in an Ecologically Oriented GIS, Proc. GIS'93 Symposium, Vancouver, Feb, 1993, pp. 167–172.
EEA, 1992, Extended Progress Note on the Preparation of Europe's Environment 1993, DG Xl Task Force, European Environment Agency, Brussels, October 1992.
KALENSKY D, 1988, Some Views on Georeferenced Digital Databases, in Building Databases for Global Science, H. Mounsey and R. Tomlinson, eds., Taylor and Francis, London, pp. 307–314.
KEUNE H and CRAIN IK, 1993, Towards the Harmonization of Environmental Measurement: Challenges and Opportunities, Proc.13th International CODATA Conference, Beijing, Oct 1923, 1992, in the press
LAM S and CRAIN IK, 1992, Representing Qualitative Attributes Using Fuzzy Data Modelling Techniques, Proc. SaskGIS Conference, Saskatoon, October, 1992, pp. 26–36.

MATHEWS JT and TUNSTALL DB, 1991, Moving Toward Eco-Development: Generating Environmental Information for Decision Makers, WRI Issues and Ideas, World Resources Institute, Washington, August, 1991.
NEWALL A, SHAW JC, and SIMON H, 1958, Elements of a Theory of Human Problem Solving, Psych. Review, May 1958, pp. 151–166.
PREMERIANE WJ, et al,1990, "An Object–oriented Relational Database", Communications ACM, Vol. 33, #11, pp. 99–108.
TOMLINSON RG, 1987, Current and Potential Use of Geographic Inforrnation Systems, The North American Experience, Int. J. G.l.S., Vol. 1, No. 3, pp. 203–218.
WCED (World Commission on Environment and Development), 1988, Our Common Future, Oxford University Press, Oxford, pp. 400.

2.5 Modell supported synthesis, evaluation and application of environmental information

H. F. Kerner

Introduction

The FAM-Project (Chapter 2.6) represents the state of the art of interdisciplinary, integrated ecosystem research – with a long term perspective, which leads to scientific ecosystem monitoring. The scientific research and monitoring of selected, representative ecosystems – and their compartments – continuously innovates and validates our understanding of ecosystem functioning and development, of their sensitivity and potentials, and of human impact on the natural resources and wildlife.

The level of integration, realized in ecosystem research projects, is not normal for many of the research and monitoring programs going on. Some investigate and monitor environmental media – water, soil, air, biota -, others select ecosystem functions, special processes or impact factors, or work on distinct levels of organisation.

The process of specialisation and differentiation of scientific disciplines continues rapidly. – In the field of basic research, this is unavoidable.

But biological and ecological systems cannot be understood and, above all, be treated, planned or developed from a sectoral point of view. The global "Man & Nature System" is an integral result of local activities, resource use, and impacts – of man and wildlife. It changes rapidly – with critical trends.

That means: The scientific community is urgently asked, to apply its knowledge – in the actual state of the art –, to make it useful for valuation, planning, regeneration, and development. We are asked to inform the public and to motify people to change their attitudes and claims in treating – themselves and the environment, consequently.

A global political consense in this field seems farer off than ever before; it may develop – forced by the attractivity and the success of a scientific conception, which is sound, valid, integral and efficient, incorporating the scientific disciplines as well, as the social, economic, and political interests.

This framework conception needs to connect global climatic models with, for instance, the mechanisms of the mineral transfer from the soil solution into plant roots. Every process, at each level of organisation, is important!

But which discipline, which program could lay down, that all research and monitoring activities should be harmonised on the working level they chose, thus making all of us look with their eyes on nature and use their terminology?!

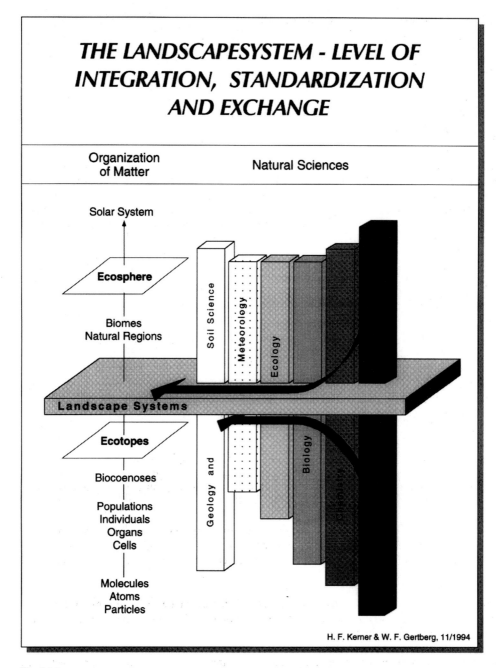

Fig. 2.5.1:
The Multidimensional, Multilevel and Multifunctional 'Building' of Sectors, Disciplines and Interests Involved in Environmental Research

Nevertheless, it's the only way out: The multidimensional, multilevel and multifunctional "building" of sectors, disciplines and interests needs an attractive, common – and commonly accepted floor – somewhat like a "belle étage" – where all disciplines and groups meet and interfere (fig. 2.5.1; it needs a common object of interest, which helps to define common working scales, joint models, data, reference values, development and quality aims – and, last not least: – which leads to a common terminolgy to make cooperation and exchange possible.

This level must be a compromise choice – anywhere in the middle of the hierarchy – between subatomic fields and the solar system – satisfying all and nobody. It should be attractive, integrative and closely connected with the planning and realisation needs.

To choice that level needs a decision – if necessary by a minority, which recognizes the actual application needs and feels responsible – being aware of things going on.

To pass now over the question: if integration and harmonisation could at all be possible – and if so: how and on what level – I will – as a provisional answer – outline a framework conception.

It developed, within the last 20 years, along with a sequence of landscape ecological projects – under the supervision of Prof. Haber – comprising the MAB 6 – Project Berchtesgaden. At last, it formed the backbone of a conception for the integral ecosystem monitoring, we are finishing at present, to be established in the German Biosphere Reserves – in the frame of the MAB-Program Sector 8 (Kerner et al. 1991; Schönthaler et al. 1994). In both cases we used a hierarchical model conception, which meets the analytical, evaluation, application and planning needs,– to identify and rank the data set to be monitored,– the mode they are needed,– their quality required, and – the sampling technic suitable.

The conception focuses on the level of ecotopes – which are, in our terminology, the puzzle pieces of the land use pattern: forest types, kinds of agriculture, types of settlements, rivers, roads and so on; they form – in a gradient between virgin, not used by man and anthropogenuous, technical (the so-called "Cultural Gradient" of ecosystems) – any Landscape System by the matter and energy exchange network they build up in space and time.

This compromise level choice – in the middle of the hierarchy (fig. 2.5.1) – is, by many reasons, excellent. Methodically (fig. 2.5.2), the central object of investigation and monitoring – the pattern of ecotopes – aggregates and integrates all sectors and levels of basic and ecosystem research into a Generalized Ecosystem Model. The model is based on the internal balance of matter and energy of the ecotopes and their external (active and passive) relations in space (fig. 2.5.3). Integrated in a space covering evaluation procedure – by using a GIS – it allows an integrated analysis and impact assessment of the environment on the landscape level – the human activities and impacts completely included.

Closely connected with the analytical "Balance-Model" is a model (fig. 2.5.4), which reproduces the environmental changes as well, as the socio-economic interdependencies within the human system – both providing the dynamic steering factors, which control the ecosystem and landscape functions and their development.

This model combination forms the core of the framework conception (fig. 2.5.5). It provides us with valid and integral information about the situation and the development trends of local

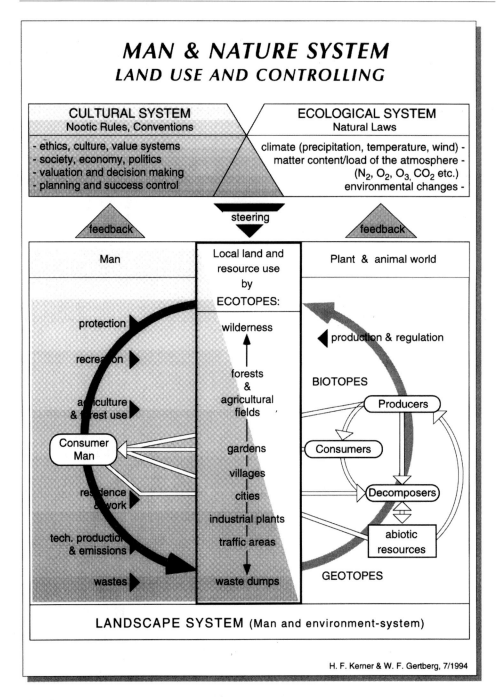

Fig. 2.5.2:
Compartements and Levels of the `Man and Nature System´

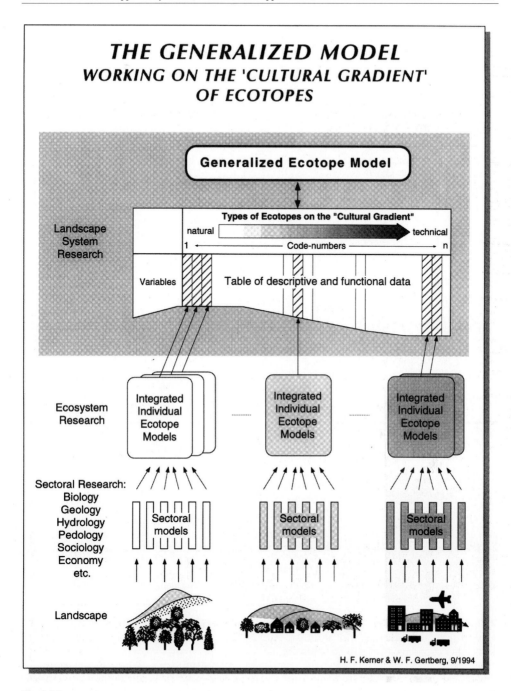

Fig. 2.5.3:
Sectoral (basic) Working Levels and Ecosystem Research Leading to the Generalized Ecotope Model to Analyze Landscape Systems

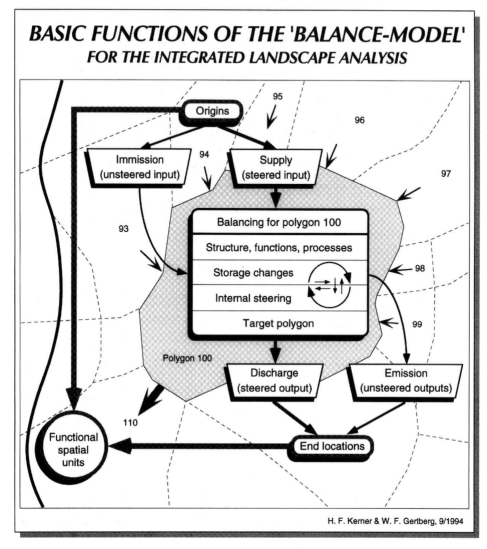

Fig. 2.5.4:
Energy, Matter and Space Functions of the 'Balance Model' Using the Basic Map of the Ecotope Pattern (1:10.000)

Man & Nature Systems – of landscape areas – in a standardized way: the same data, the same evaluation results; that is: with information, which is directly comparable and exchangable whereever the instrument will be applied.

The global picture of state and trends of the Biosphere (fig. 2.5.6) is an extrapolation – model based; result of computer aided aggregation steps, which transport the local (landscape) research, monitoring, and evaluation results up to the global level. The more – compatible – monitoring and research stations we have, in a global network, the better the quality and validity of the global analysis will be.

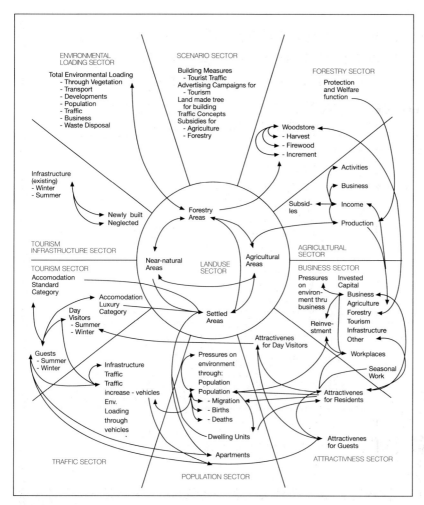

Fig. 2.5.5:
The 'REGIO' Model Conception as an Example of Simulation of the Cultural System (from Grossmann in KERNER et al. 1991)

That's the proposal – integral and integrative, applicable for the environmental analysis and planning, the environmental impact assessment as well, as for the interpretation and evaluation of monitoring data and the selection of measures of regeneration and development. The conception is capable of development – by evolution and innovation through basic research and along with the progress of methodical developments.

It answers the question, if the integration of the sectoral diversity and the harmonisation of working levels and scales may be possible: They are – methodically and in practice – and any, even critical discussions about the solution offered here – may contribute at least to its improvement.

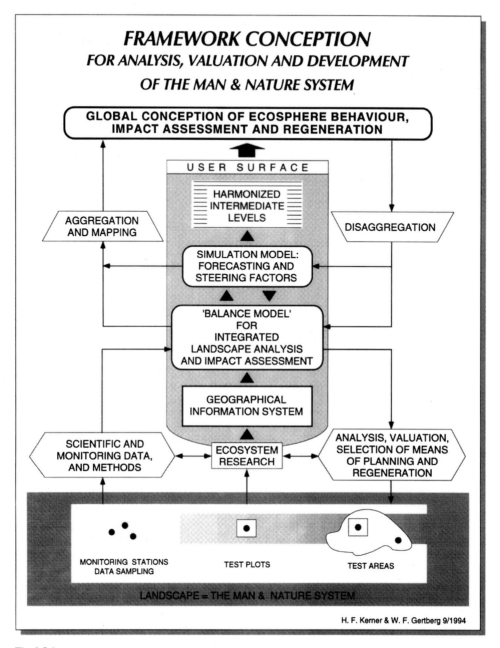

Fig. 2.5.6:
A Hierarchical Framework Conception for Integration and Aggregation of Basic Ecological Research Levels for Global Ecosphere Assessment and Simulation

References

KERNER H.F., SPANDAU L. and KÖPPEL J.G. (1991). Methoden zur Angewandten Ökosystemforschung, entwickelt im MAB 6 projekt: Ökosystemforschung Berchtesgaden 1981–1991. In: Abschlußbericht Projektsteuerungsgruppe MAB-Mitteilungen 35.1 und 35.2, Bonn.
SCHÖNTHALER K., KERNER H.F., KÖPPEL J.G. and SPANDAU L. (1994) Konzeption einer ökosystemaren Umweltbeobachtung in Biosphärenreservaten. Pilotprojekt für Biosphärenreservate. Umweltbundesamt, Berlin

2.6 Integrated assessment concept of the working group "Informationsystems for environmental research and planning" at the GSF

R. J. M. Lenz, M. Knorrenschild, M. Schmitt and R. Stary

Introduction

In 1993, a Working Group on Information Systems for Environmental Research and Planning was established as part of the project group on potential threat to the environment posed by chemicals (PUC, Projektgruppe Umweltgefährdungspotentiale von Chemikalien) at the GSF. In addition to the tasks of PUC; it focuses on environmental impacts by various management practices and use of non-synthetic substances like nitrogen, and specifically the storage and retrieval of data and models relevant for impact assessment studies on the ecosystem and landscape level. The basis for this work are databases including Geographical Information System (GIS), and models, as well as integration methods like (geo)statistics, and environmental classification. Experience gained by case studies, integration and classification approaches for ecological systems will have a feedback on the basic information systems and integration methods.

The group is part of the research network on agroecosystems in Munich (FAM, Forschungsverbund Agrarökosysteme München) and collaborates closely with various GSF-institutes such as those for Soil Ecology and Ecological Chemistry. The collaboration includes important input of experimental data, eco-statistics and modeling. The computing centre of the GSF provides the group with access to a mainframe computer and database management systems.

Two main projects and one case study from the working group are described briefly in the following.

The FAM Database

In 1990, agricultural production at the research farm of Scheyern (Bavaria) was changed from conventional methods to an integrated and ecological cultivation. Several hundred environmental parameters will be continuously measured over a period of 7–15 years at about 400 relevees in a 50 m grid. This large amount of data is being stored in a relational database, which itself will be extended in four stages to provide a continuous feedback and the basis for integrated evaluations of the project (Fig. 2.6.1, [1]). The group is concerned with the development of this database.

- *Stage 1*: Measured data from agriculture, soil science, hydrology, meteorology, geobotany and other ecologically relevant disciplines will be entered into the database and related to spatial data processed in the GIS. The GIS also includes data from more extensive detection and assessment in the surrounding area of the research farm (550 square kilometers, map-scale 1:25.000).

- *Stage 2*: Meta information, e.g. data formats and parameter units, methods of measurement and sample analysis, will be included in the FAM database.

- *Stage 3*: A graphical user interface will be implemented to facilitate access to preprocessed and raw data for members of the different working groups. Thus the database can be used by scientists of various disciplines to obtain input data for geostatistical analysis, GIS, and modeling. In addition, interfaces to graphics, spreadsheets and word processing applications will be available for report generation. External users will have access to released data via a retrieval system on the basis of the World Wide Web computer programme (WWW) and mosaic.

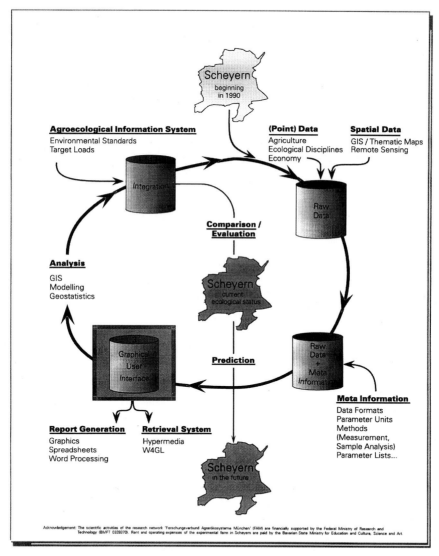

Fig. 2.6.1:
The Scheme for the FAM Relational Database.

- *Stage 4*: External data like environmental standards and target loads from the Agricultural Information System (AIS) will be integrated into the database for comparison and evaluation of the current ecological status of Scheyern. Relevant parameters will be extracted from the data pool and correlated with spatial relation patterns to enable prediction of the ecosystems status and environmental risks in Scheyern and similar agroecosystems.

UFIS – a database of ecological models

Introduction

Due to increasing activities in ecological modeling over the past decades, it has become difficult for the individual ecologist to stay informed about the state of the art. While modellers usually are in touch with others working on the same subject, there is often lack of knowledge outside their own specialty. There is also the problem of data relevant to models: input data required by modelers are not necessarily the data that are acquired by field ecologists. The scientific community is already familiar with databases that offer information and access to various kinds of data; however, we are not aware of a database on models that allows various applications while at the same time giving detailed information e. g. on process formulations.

A database of ecological models would contribute greatly to an efficient modeling process. Such a database is crucial to international coordination of modeling activities. Model comparisons and questions of model coupling have been defined as key goals in many activities like, for example, IGBP (International Geosphere-Biosphere Program; [1]). It is the mission of UFIS to establish such an information system (UFIS is the German acronym for environmental research information system). Work on the UFIS project started in autumn 1992 at the GSF-Research Centre for Environment and Health and is being funded by the German Federal

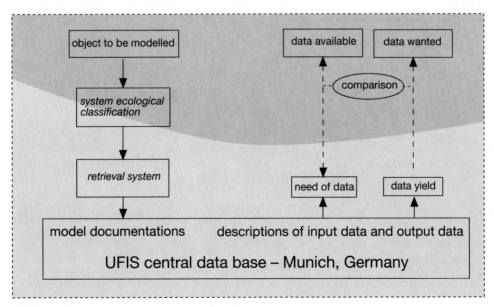

Fig. 2.6.2:
The UFIS Project: Database with (online) Information on Ecological Models

Ministry of Research and Technology. UFIS is designed to offer information on ecological models and their data requirements to a degree of detail required to evaluate the quality of a model in several respects (Fig. 2.6.2). In view of IGBP relevance, UFIS will focus in particular on ecosystem models (while not neglecting models at lower levels of organization). In the following the main features of UFIS and its organization are described [3].

Users of UFIS

Individual scientists of various backgrounds will take advantage of UFIS. With the help of UFIS, field ecologists will be able to identify present gaps in available data that prevent full use of state of the art models. On the other hand modelers will be able to get an overview of models available at other institutions in order to avoid duplication of efforts. Typical inquiries may be asking for a list of all models that simulate the water balance within an agricultural ecosystem, or for models that yield the biomass of an individual as an output variable. Modellers will also be able to extract process formulations from models in order to combine them with their own. The information provided will carry particulars about the temporal and spatial range in which models yield valid results. This is particularly important to scientists with a different speciality who may not be able to judge if some model meets their goals. The amount of detailed information on the models stored in UFIS ensures that models can be compared to each other in respect of different formulation of processes, different data requirements etc. This makes UFIS a valuable tool for project managers in international modeling activities like IGBP.

What kind of information is stored in UFIS?

A first concept of a documentation scheme for ecological models has already been proposed by Asshoff of the Ecosystem Research Centre at the University of Kiel [4] (cf. Chapter 2.2). In joint work with Asshoff and Benz of the University of Kassel, we have developed a complete documentation scheme. The information contained in this scheme on a particular model is threefold: general information, description of processes within the model, and technical information.

General information
The aim of this section is to provide information for the user who is not interested in details of the model. Besides the obvious items like name and address of the modeller, a brief summary of the aim of the model and its basic ideas, valuable information regarding the range of applications is given. This includes references to other models that could possibly be coupled with the one under consideration. The suitability of the model for purposes other than the original one can to a certain extent be judged from reports on calibrations and validations performed (and the data used) as well as spatial and temporal scales. References to user manuals and on detailed model documentation are also available. A flow diagram of the model structure will provide the UFIS user with an idea of the organization of the model and its submodels.

Description of processes within the model
The basic idea is that each model is composed out of a number of hierarchically ordered processes. Information is stored for each process separately. It includes: brief description, mathematical formulation, possible constraints, initial/boundary conditions, spatial and temporal resolution, references to the literature and embedded subprocesses. Quantities used in the mathematical formulation will be classified as input and output for each process, indicating al-

so which quantities are input and output with respect to the full model. For each of the quantities a brief description, values used, and range of values possible, is given and also the source of the data (where applicable). For each single process a table of environmental conditions contains a description of its real world setting. Entries in this table describe the physical environment in which the underlying process formulation is valid. One of the items asked is the "level of organization" ([5], [6], [7]) on which the process takes place and a more specific characterization (which type of ecosystem, which species etc.). Checking lists for these specifics are supplied; in particular a comprehensive thesaurus for types of ecosystems has been developed. Further items asked for are geographic location, type of climate, type of soil. Where meaningful and available, information will also be given by means of commonly used classification schemes (like, for example, the FAO scheme for soil types).

Technical information on a model
This part deals with the computer implementation of the model. It is divided in software- and hardware related information. Under the topic "software", details are given on which libraries are needed to run the model (if any), programming language used and where to get a complete program documentation. For the goal of UFIS, as outlined above, the data question is of particular importance. The UFIS user will find here detailed descriptions of the input files needed and also where to obtain them from. Sources of the program code itself, demo input, and sample output files are also given. Hardware requirements are given in the second part of this section, i. e. which operating systems versions of the model codes are available for, as well as disk space required.

Technical realization of UFIS
In view of the amount of paperwork required by the modeller who wants to (or must) provide the full documentation of his/her model work, a computerized version of the questionnaire is in progress. This program will guide the modeller through all the questions being asked, and will accept answers on screen. Entries accepted will be saved to a file which is then sent to the UFIS managers where it will be double-checked for completeness, unique use of terms etc. and finally stored in the central UFIS database. UFIS will be implemented using the INGRES relational database management system and will be maintained on a CONVEX computer at the GSF research centre. It will be accessible to the scientific community via Internet and WWW in order to enable queries via remote login to the database server and to yield immediate search responses.

Concluding remarks

The UFIS questionnaire described above is available (in English or German) from the authors upon request. It comes with detailed instructions on how it is to be filled out as well as a completed sample questionnaire. We are currently in the process of feeding completed questionnaires into the database. At present, there does not seem to be any internationally standardized model documentation available elsewhere and it is hoped that UFIS will fill this gap.

A case study in integrated assessment: landscape diversity and land use planning in the surrounding of the research farm Scheyern

Introduction

A maximum of landscape diversity is to be regarded as a helpful target as well as an analytical tool to detect the present planning deficits. First of all, specific indicators and targets for ecosystem and landscape levels will be defined. Ecosystem types are derived from landscape ecological site conditions, and are combined with the demands of minimum areas related to present and potential ecosystem types.

The suitability of management practices for the site conditions, and the requirements mentioned above, will jointly form the basis on which a target landscape in its pattern, and the management practices suggested can be mapped. In comparison with the present use of the landscape, measures are derived which indicate that the target of a high landscape diversity has real practical meaning: conversion of about 15% of the arable land into forests and grasslands as one result of this study fits in to the political and administrative programs already established. In addition, four modification categories of the actual usage are mapped and described in size and number in this case study. The basic assumptions of this approach will be made transparent, so as to provide further working hypotheses for an interdisciplinary and integrative research into landscape ecology in rural landscapes ([8]).

Material and Methods

The test area chosen for this study is a mapsheet with a scale of 1:25.000, TK 25 no. 7435 Pfaffenhofen, in the tertiary hills of Bavaria, about 50 km north of Munich. The altitude ranges from 400 to 525 m above sea level; the average annual temperature is 7–8° C. Rainfall is high during the summer with mean daily amounts of about 50 mm every two years. High relief energy in combination with highly erosive soils, and a complex groundwater system are naturally important conditions for the sensitivity of the landscapes' abiotic resources. Due to the intensive agricultural landuse during the past centuries, especially during the last decades, only few and small ecologically important sites have survived. Soil classes, land use types, specific climatic conditions like danger of frost in spring or autumn, ecologically important sites (mapped biotopes), aspect and slope, depth of the groundwater level etc. are basic information which characterize both the ecosystems and the suitability of management practices. The data are processed in GIS, and some results will be presented in the following sections.

Results

As a result of the comparison 'existing with potential', which is synonymous with present and potential land use types, are described in Figure 2.6.3 and 2.6.4. The spatial fixation and description of the results is shown in Figure 2.6.5. Three general measures can be distingui-shed:

(i)   modification of forestry (ecosystem level): 21.2% (of the total area of the mapsheet)
(ii)   modification of arable land (44.5%) and grassland (10.7%) (ecosystem level)
(iii)   conversion of arable land to forests (6.5%) or grasslands (7.5%) (landscape level)

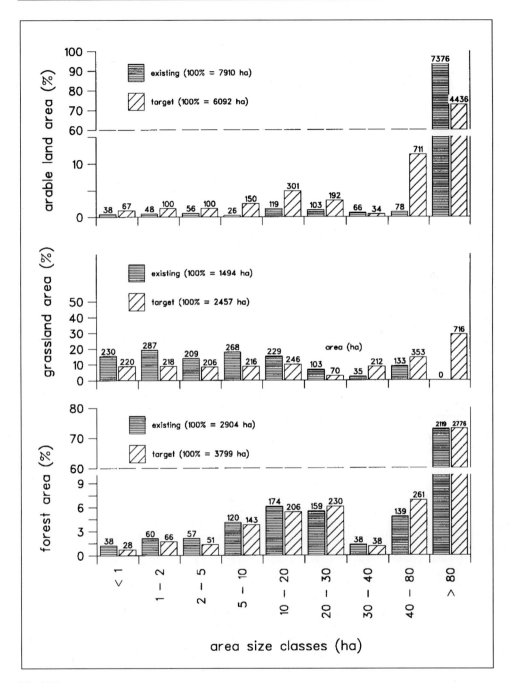

Fig. 2.6.3:
Comparison of Existing and Potential Land Use Types in Terms of Area Size Classes.

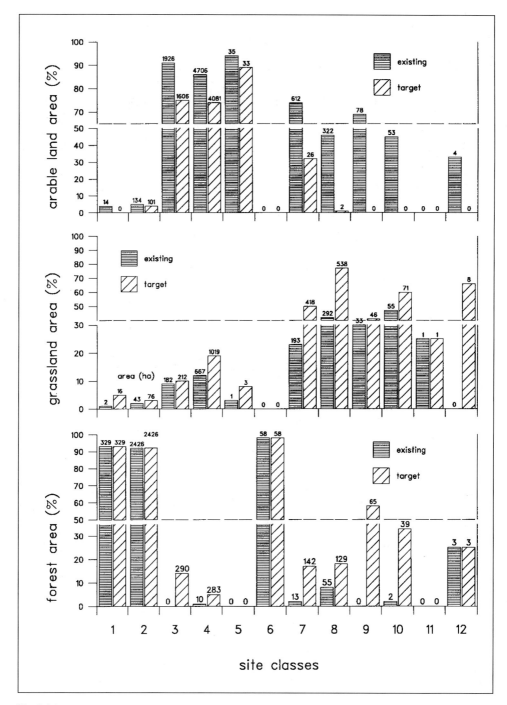

Fig. 2.6.4:
Comparison of Existing and Potential Land Use Types in Terms of Site Classes

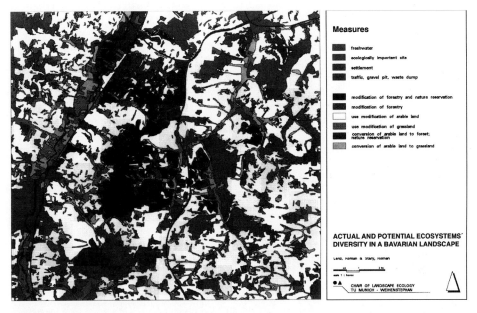

Fig. 2.6.5:
Landscape Diversity and Land Use Planning: The Measures Needed to Reach the Target of Highest Possible Ecosystem Diversity in the Research Area.

Modifications of land use practices to more conservation and extensification as suggested in Figure 2.6.3 and 2.6.4 in more detail will soon only be limited by the question of land ownership. Convincing owners of land, by proper compensation, or establishing socioeconomically appropriate land use practices, are the task of the future.

To summarize, about 15% (1210 ha) of the existing arable land is situated on sensitive and/or ecologically improbable sites, so that resource protection and optimization of diversity with arable farming is not possible – the usage of these sites has to be converted to forest or grassland. Hence it is suggested to enlarge the forested areas from 21% to 28%, the grasslands from 11% to 18%, and to reduce the arable land from 58% to 44% of the total areas in this region under study.

Discussion

The more idealistic the approach the more practical the results: 15% conversion of arable land, and furthermore the modification of forests, arable land and grassland are in keeping with the programs of the set aside policy and the nature protection programs. In the municipality of Rudelzhausen, for instance, in 1993 about 18% set aside areas could be mapped. Nevertheless, realization of the modification measures will be difficult due to the necessity to convince and compensate the owners of the areas according to the targets and programs. The general framework conditions to conduct the measures are better than ever, and to have a simple but ecologically sound and integrative target for a landscape is most important to improve landscape planning.

Conclusions

Data and GIS models and evaluations are the basis of the information systems for the purpose of integrated assessments, as in the case study on landscape diversity. What could not be shown in this contribution, although very essential for the transparency and validity of integration methods are (i) the statistical analysis of data (geostatistics, "error maps"), and (ii) the classification of the ecological systems under observation. The issue of clear descriptions of the object is a striking one, not only because of the greatness and complexity of the environment, but also to ensure that the same subject is dealt with, for example, while comparing methods. Otherwise confusion, often already initiated by different and unclear targets, will occur. Hence, it is the task of the scientific community to provide politics and society with a better understanding and description of our ecological systems, and the task of the society to define and to explain targets.

Figure 2.6.6 shows the overall situation in dealing with integrated assessment. The two worlds – science (quantitative, causal) and technology or management (qualitative, empirical) have to be bound together by common theories and models, based on scientific experiments and expe-

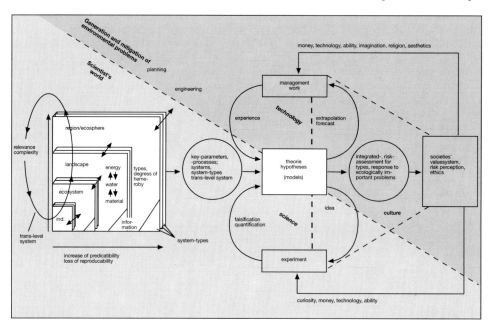

Fig. 2.6.6:
Environmental Systems and Theories in a Transdisciplinary Context.

rience in management, and brought into the societies value system. To improve civilization and culture as an ecologist, this means to describe ecological data, methods and objects clearly and to prepare the confrontation best by linking experimental data with management work, and both of them with the societies value system. A (small) working group has to deal with selected methods and some case studies, but with a better harmonization and standardization of environmental data and models in the broad scientific community we do have some more chances to fill up a little more the conceptual and theoretical approaches.

References

[1] SCHMITT M., LENZ R. J. M., WENDEROTH Chr., PEPLOW H., 1994: Structure, contents and perspectives of the FAM database. Poster at the TERN conference in March 1994 at the GSF, Neuherberg, Germany (unpublished).
[2] IGBP Global Modelling and Data Activities 1994 – 1998. IGBP Report 30, Stockholm 1994
[3] KNORRENSCHILD M., LENZ R.J.M., FORSTER E.-M., HERDERICH C., 1993: UFIS – a database of ecological models. Proceedings of the International Congress on Modelling and Simulation, Dec. 1993, Perth, Australia (ed. Mc Aleer, Jakeman), vol. 1, pp. 379 – 382
[4] M. ASSHOFF: Konzept einer Modelldokumentation. Interne Mitteilungen 5, Projektzentrum Ökosystemforschung, Univ. Kiel, 1991.
[5] G. T. MILLER: Living in the Environment – Concepts, Problems, and Alternatives. Wadsworth Publ., 1975.
[6] LENZ, R.: Systemökologische Anforderungen an ein Umweltforschungsinformations-system (UFIS) als operationale Basis für Scaling – Konzepte in der Umweltplanung. In A. Jaeschke et al., editor, Informatik für den Umweltschutz, pages 427 – 440. Springer Verlag, 1993.
[7] LENZ, R.: Ecosystems Classification By Budget of Material – the Example of Forest Ecosystems Classified as Proton Budget Types. In: J. Klijn (ed), Ecosystem Classification for Environmental Management, Kluwer Academic Publisher, 117 – 137 (1994)
[8] LENZ R. J. M., STARY R., in press: landscape diversity and land use planning: a case study in Bavaria. Urban and Landscape Planning.

2.7 Estimating vertical fluxes of gases between atmosphere and ecosystems

U. Dämmgen, L. Grünhage and H.-J. Jäger

Introduction

Natural and man made terrestrial and aquatic ecosystems have been affected by air-borne matter which may be regarded both as nutrient or as pollutant. Much attention is being paid to the effect of increasing CO_2 concentrations which might constitute a fertilizing effect. Unwanted fertilizing effects have been attributed to depositions of ammonia and nitric oxide. The decrease of depositions of sulphur both from sulphate containing particles and from sulphur dioxide is being made responsible for sulphur deficiency in European agriculture. In connection with global change not only increased CO_2 concentrations, but also increased O_3 and SO_2 concentrations are being discussed. Their effects are thought to be of commercial and ecological importance. However, global and regional models are not able to determine inputs of airborne matter into ecosystems with an adequate accuracy due to the fact that only concentrations of the respective species are taken into account.

The following discussion aims to provide the background information necessary, to present comparatively simple models describing flux densities of air-borne matter, and recommends measures to meet the needs for the application of these models for monitoring purposes.

Vertical fluxes as doses in dose-response relationships

Classical theory in toxicology provides an agonist receptor model for the derivation of basic understanding of dose-response mechanisms. According to this model the effect of an active substance (pharmacon) on an organism is a function of its concentration at the site of the respective receptors within the organism, which we consider to be the effective stressor dose (Dämmgen et al. 1995). In the steady state this concentration is a function of the flux of the active substance to the site; the flux from the site is a function of the concentration itself and depending on the ability of the organism to detoxify the substance by excretion, metabolisation, dilution or immobilisation of the active substance (e.g. Kurz et al. 1990).

Measurements of the effective concentrations of the active substance at the site of the receptor are possible within the scale of cells or organs, if at all; they are impossible within the scale of individuals or ecosystems. Therefore, the best approach towards an establishment of dose-response relationships is the quantification of the entity governing the effective concentration, i.e. the flux into the system of the active substance or its precursor, as the active substance is not necessarily the substance absorbed by the organism or system. This flux of the potential stressor into the system is called the absorbed stressor dose. However, detoxification fluxes, which happen during transport inside the system reducing the effective stressor dose, cannot be quantified either.

Therefore, the only entity which can be measured, is the absorbed stressor dose. In case, the stressed system is an (extensive) ecosystem and the stressor an air-borne species, the absorbed stressor dose has to be determined as vertical flux density, i.e. a quantity of matter per unit of area and per unit of time (Dämmgen et al. 1995).

Thus it is sensible to relate any potential impacts of air-borne matter to ecosystems, either as pollutant (O_3, SO_2, NO_2, NH_3) or as nutrient (CO_2, NO_2, NH_3) to the respective vertical flux densities.

Downward vertical fluxes of trace gases are normally not proportional to their concentration in ambient air. They are dominated by the vertical exchange properties of the atmosphere near the ground and by the ability of the system to absorb these gases (Roden and Pearcy 1993; McMurtrie and Wang 1993; Grünhage and Jäger 1994a).

Dose-response relationships are normally established by using closed or open-top chambers. In this case the use of concentrations or weighted concentrations is adequate and gives good results, as the vertical exchange properties of the air are constant. The results obtained, though, cannot be transferred to field conditions (Krupa et al. 1994); the use of concentrations only does not lead to unambiguous dose-response relationships: high concentrations combined with small turbulences can have the same toxicological effect as medium concentrations with medium turbulence (Bugter and Tonneijck 1990; Tonneijck and Bugter 1991) and can result in the same flux densities towards the canopy (Grünhage & Jäger 1994a).

The experimental determination of vertical flux densities

Thermal energy (sensible heat and latent heat), mechanical momentum and matter are transported effectively in the air by turbulent diffusion only. Any transport can be understood as a flux along a gradient of potentials P along z which is made possible by the specific conductivity of the transporting medium. Fluxes between two potentials into a horizontal plane A along a distance z can thus be expressed as

$$\Phi = -(P_1 - P_2) \cdot \kappa \cdot \frac{A}{z} \tag{1}$$

with P_1 upper potential at z_1
P_2 lower potential at z_2

$$F_c = -(P_1 - P_2) \cdot \kappa \cdot \frac{1}{z} \tag{2}$$

Equation (1) is analogue to Ohm's law and is fundamental with respect to the resistivity models used to describe vertical flux densities within a given medium, i.e. in air or in plants (Thom 1975; Fowler & Unsworth 1979). For the exchange of matter between atmosphere and ecosystems it is usual to deal with flux densities F rather than fluxes:

Prerequisite for the use of this equation in air is that the vertical flux density F_c of matter is independent of height: Fc can be determined above the canopy, whenever the fetch is adequate and all measuring heights are within the constant flux layer and whenever sinks or sources of for the respective species are absent in this layer. This can be assumed to be valid for the constant flux layer for momentum, sensible and latent heat and for inert gases. Corrections have to be made for reactive species. The vertical extension of the constant flux layer is restricted by the fetch available and by the physical properties of the canopy.

Thus, the determination of any fluxes presupposes the measurement or modelling of the two potentials and of the specific conductivity of air. It is normally assumed that the potentials governing matter transport are the concentration gradients a of the respective species, and that the conductivity can be derived from the conductivies for sensible heat and momentum.

$$F_c = F_H \cdot \frac{p_a(z_1) - p_a(z_2)}{\Theta(z_1) - \Theta(z_2)} \tag{3}$$

or, if $\Delta\Theta \approx 0$,

$$F_c = F_M \cdot \frac{p_a(z_1) - p_a(z_2)}{u(z_1) - u(z_2)} \tag{4}$$

If these can be measured directly – for instance by use of sonic anemometers (Grünhage et al. 1994) –, flux densities of matter can be determined as

with
F_H flux density of sensible heat
$p_a(z)$ mass concentration (partial pressure) in the atmosphere at height z
$\Theta_{(z)}$ potential air temperature at height z
F_M flux density of momentum
$u_{(z)}$ mean horizontal wind velocity at height z

If F_H and F_M cannot be measured directly, they can be derived from the gradients of temperature und horizontal wind velocity (cf. Etling 1987).

All measurements require sophisticated and expensive equipment and experienced personell. Flux density measurements are therefore restricted to few places. Continuous measurements are rarely to be found. Passive sampling techniques so far do not normally provide an alternative (Dämmgen et al. 1995).

Estimating vertical fluxes of CO_2 and trace gases from concentration data

Concentrations ρ of trace gases have been monitored by autorities mainly for the sake of public (human) health. Several approaches have been made to make use of these data to estimate dry depositions of trace gases and aerosols. They all make use of the resistance analogy provided by eq. (5):

$$F = \frac{p_a(z) - p_b}{R_a(z)} = -\frac{p_b - p_s}{R_b} = -\frac{p_s - p_b}{R_c}$$

$$= -\frac{p_a(z) - p_i}{R_a(z) + R_b + R_c} \tag{5}$$

with
$\rho_a(z)$ concentration in the atmosphere at a height z
ρ_b concentration just outside the laminar boundary layer
ρ_s concentration at the leaf surface
ρ_i concentration inside the plant
R_a atmospheric (columnar) resistance due to turbulent diffusion
R_b laminar boundary layer (columnar) resistance due to molecular diffusion
R_c effective bulk canopy (columnar) resistance including all resistances within the plant

For a comprehensive "curcuit diagram" cf. Hicks et al. (1987). Negative signs denote fluxes into the plant.

Constant mean deposition velocities

The easiest way to estimate downward vertical flux densities (depositions) is by application of the so-called deposition velocity vD which relates flux densities of a species A to the respective concentrations via:

$$F_A = -\ ^vDa \cdot \rho A \tag{6}$$

Comparison of eqs. 1, 5 and 6 reveals that this assumption can be made if $P2$ (sink potential, ρi) is negligible with regard to P_1 (source potential, $\rho_a(z)$). Furthermore it presupposes that a mean concentration can be combined with a mean deposition velocity in such a way that does not contradict Reynold´s averaging rules.

These limitations are almost never met. For many gases of interest, in particular CO_2 or reactive nitrogen species, potentials P_2 are not negligible. Both potentials exhibit annual and diurnal patterns. Also, atmospheric conductivities and therefore deposition velocities have distinctive annual and diurnal patterns (Fig. 2.7.1 and 2.7.2).

Numerous lists of deposition velocities have been published for different sinks (vegetation, soil, water surfaces, etc.) under different conditions. They vary extraordinarily. Most of these lists do not provide the necessary information needed for their transfer.

The use of deposition velocities is restricted to rough estimations of the order of magnitude of a potential deposition for those gases which flow downward only (O_3, HNO_3). Deposition velocities are useless for dose-response considerations.

In a changing climate deposition velocities will change, because the exchange properties of the air will change due to changed wind velocities and changed temperatures; also the sink properties of the canopy are liable to change.

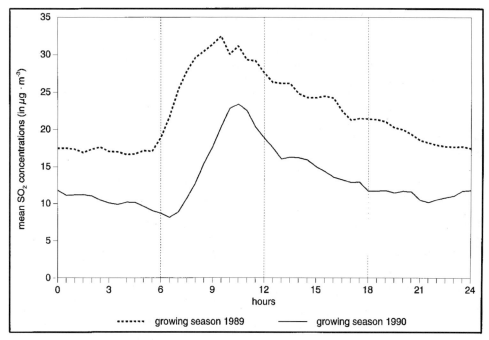

Fig. 2.7.1:
Mean Diurnal Patterns of SO$_2$ Concentrations (adapted from Grünhage et al. 1992)

Variable deposition velocities – inferential method

The disadvantage of the vD approach may be overcome by estimating the deposition velocity from the atmospheric, diffusive boundary and canopy resistivities (R_a, R_b and R_c):

$$V_D = \frac{1}{R_a} + \frac{1}{R_b} + \frac{1}{R_c} \tag{7}$$

Hicks et al. (1987) described a method to obtain the parameters needed from measurements, the applicabilty was examined in detail by Matt & Meyers (1993). The computation from field data of the respective resistivities as described by Hicks et al. (1987) reveals that measurements above and in real ecosystems are needed to provide the information about the fluxes in these systems. The resistances computed from these data are then used to infer the fluxes into similar systems.

Variables to be determined (15 minute means) are:
– source concentration ρ at standard height
– global radiation G at standard height
– horizontal wind velocity u at standard height
– standard deviation of wind direction (10 s intervals)
– air temperature T at standard height
– relative humidity
– surface wetness

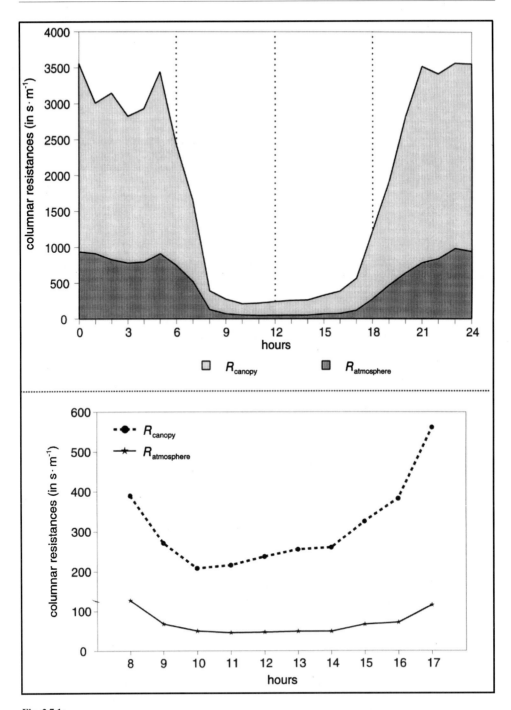

Fig. 2.7.1:
Mean Diurnal Patterns of Atmospheric and Canopy Columnar Resistances (adapted from Grünhage & Jäger 1994a)

Maximum absorbed doses - exposition potential of the atmosphere

Grünhage & Jäger (1994a) transformed eq. 5 assuming that $P_2\ (\rho_1) = 0$ and that R_b and R_c can be summarized to R_{Plant}:

$$F = \frac{\rho_a(z)}{R_a(z) + R_b + R_c} = + \frac{\rho_a(z)}{R_a(z) + R_{Plant}} \tag{8}$$

With

$$\frac{1}{R_a(z)} = \lambda(z) \tag{9}$$

and

$$R_{plant} = 0 \tag{10}$$

a maximum flux towards the canopy can be calculated

$$F = -\rho_a(z) \cdot \lambda(z) = -E(z) \tag{11}$$

The product $\rho_a(z) \cdot \lambda(z) = E(z)$ can be regarded as the exposure potential of the atmosphere. In reality, the absorbed dose is smaller than $E(z)$, because $R_{Plant} > 0$.

Whereas R_{Plant} is a function of the momentary state of the canopy depending on plant species and activity, E governed by the turbulent exchange of the atmosphere and the (physical) properties of the canopy (roughness) influencing the shearing stress.

E has been modelled from routine data for short vegetation (Grünhage & Jäger 1994b) from the following variables (half-hourly means):

- source concentration ρ at standard height
- global radiation G at standard height
- horizontal wind velocity u at standard height
- air temperature T at standard height

Absorbed doses

The exposure potential $E(z)$ is related to the "real" absorbed dose by a correction factor S (sink strenght, $0 < S < 1$) according to

$$F = -E(z) \cdot S \tag{12}$$

The determination of S still requires flux density measurements, from which models can be derived. Grünhage & Jäger (1992) suggested a model for S based on CO_2 flux densities. Calibrated models should be useful to provide the information needed to transfer data from one measuring site to other sites or other types of vegetation. Plant growth models should in principle be suitable to contribute to the determination of S for important species in the respective regions.

Levying of environmental data to estimate vertical fluxes

Agriculture and natural ecosystems are affected by depositions of air-borne matter. The extent to which this happens is not reflected by mean concentrations and mean annual flux densities, but by patterns of flux densities which have to be determined with a comparatively high resolution in time (days rather than weeks). To establish adequate data, (contiuous) flux density measurements will have to be performed at suitable sites in the regions potentially affected for all systems of major interest. From these measurements data may be inferred on different ecosystems by providing adequate monitoring facilities including not only the concentrations in ambient air but also the means to estimate fluxes with an appropriate accuracy.

Nevertheless, the development of adequate means to measure flux densities by cheap and simple equipment (passive monitoring of flux densities) is important at least for those air-borne species which contribute to element balances.

References

BUGTER, R. J. F.; TONNEIJCK, A .E .G. (1990): Zichtbare beschadiging bij indicatorplanten in relatie tot ozon en zwaveldioxide. Report no. R-89-10. Centre for Agrobiological Research, Institute for Plant Protection, Wageningen.
Dämmgen, U.; Grünhage, L.; Haenel, H.-D.; Jäger, H.-J. (1993): Climate and Stress in Ecotoxicology. A Coherent System of Definitions and Terms. Angew. Botanik 67, 157–162.
DÄMMGEN, U.; GRÜNHAGE, L.; JÄGER, H. -D.; HAENEL, H. -D. (1995): Model Considerations for the Use of Passive Sampling Techniques to Determine Trace Gas Concentrations and Flux Densities. Water Air Soil Pollut., in press.
ETLING, D. (1987): The Planetary Boundary Layer PBL. In: Landoldt-Börnstein. New Series. Group IV, vol. 4., Meteorology. Subvil. c, Climatology. Part 1. Springer, Berlin. pp. 151–188.
FOWLER, D.; UNSWORTH, M. H. (1979): Turbulent transfer of sulphur dioxide to a wheat crop. Quart. J. R. Met. Soc. 105, 767–783.
GRÜNHAGE, L.; DÄMMGEN, U.; JÄGER, H.-J. (1992): Das chemische Klima (I): Konzentrationen in der Umgebungsluft. in: Grünhage, L.; Jäger, H.-J. (eds): Auswirkungen luftgetragener Stoffe auf ein Grünlandökosystem – Ergebnisse siebenjähriger Ökosystemforschung – Teil I. Landbauforschung Völkenrode, Special Issue 128, 135–187.
GRÜNHAGE, L.; DÄMMGEN, L.; HAENEL, H. -D.; JÄGER, H. -J. (1994): Response of a grassland ecosystem to air pollutants. III. The chemical climate: Vertical flux densities of gaseous species in the atmosphere near the ground. Environ. Pollut. 85, 43–49.
GRÜNHAGE, L.; JÄGER, H. -J. (1992): Ableitung einer "flußorientierten" Kenngröße für die Festlegung von Grenz- und Richtwerten zum Schutz der Vegetation – critical doses. in: Grünhage, L.; JÄGER, H. -J. (eds): Auswirkungen luftgetragener Stoffe auf ein Grünlandökosystem – Ergebnisse siebenjähriger Ökosystemforschung. Teil I. Landbauforschung Völkenrode. Special Issue 128, 253–263.
GRÜNHAGE, L.; JÄGER, H. -J. (1994a): Influence of the atmospheric conductivity on the ozone exposure of plants under ambient conditions: considerations for establishing ozone standards to protect vegetation. Environ. Pollut. 85, 125–129.
GRÜNHAGE, L.; JÄGER, H. -J. (1994b): Atmospheric ozone exposure-potential for vegetation: how suitable are critical levels? In: Fuhrer, J.; Achermann, B. (eds.) Critical Levels for Ozone. A UN-ECE workshop report. Schriftenreihe FAC Liebefeld 16, 222–230.
HICKS, B.B.; BALDOCCHI, MEYERS, T.P.; HOSKER, R.P.; MATT, D.R. (1987): A preliminary multiple resistance routine for deriving dry deposition velocities from measured quantities. Water Air Soil Pollut. 36, 311–330.
KRUPA, S. V.; GRÜNHAGE, L.; JÄGER, H. -J.; NOSAL, M.; MANNING, W.J.; LEGGE, A.H.; HANEWALD, K. (1994): Ambient ozone (O_3) and adverse crop response: a unified view of cause and effect. Environmental Pollution (in press)
KURZ, H.; NEUMANN, H.-G.; WOLLENBERG, P.; FORTH, W.; HENSCHLER, D.; RUMMEL, W. (1990): Allgemeine Pharmakologie. In: Forth, W.; Henschler, D.; Rummel, W. (eds): Allgemeine und spezielle Pharmakologie und Toxikologie. Bibliographisches Institut & Brockhaus, Mannheim. S. 1–97.

MATT, D. R.; MEYERS, T.P. (1993): On the use of the inferential technique to estimate dry deposition of SO_2. Atmospheric Environ. 27A, 493–501.
McMURTRIE, R.E.; WANG, Y.P. (1993): Mathemaical models of the photosynthetic response of treestands to rising CO_2 concentrations and temperatures. Plant Cell Environment 16, 1–13.
RODEN, J.S.; PEARCY, R. W. (1993): The effect of leaf flutter on the flux of CO_2 in poplar leaves. Functional Ecology 7, 669–675.
THOM, A.S. (1975): Momentum, mass and heat exchange of plant communities. In: Monteith, J. L. (ed.): Vegetation and the Atmosphere. Vol. I. Principles. Academic Press, London. pp 57–109.
TONNEIJCK, A.E.G.; Bugter, R.J.F. (1991): Biological monitoring of ozone effects on indicator plants in the Netherlands: Initial research on exposure-response functions. VDI-Berichte 901, 613–624.

2.8 Environmental-economic accounting: how can it support decision-making

W. Radermacher

Objectives

Nature has much to offer for economic use. It supplies energy and raw materials, provides the location for businesses, and serves as medium receiving pollutants, waste, etc. However, its potential is not unlimited, but it is reduced, at worst even destroyed, by being used. The fact that *nature is a factor of production* whose scarcity must be considered in economic accounting has been noted only in recent years as environmental problems are becoming more serious and pressing.

The statistical coverage of changes in "nature capital" due to economic activities is what Environmental-Economics Accounting aims at. The idea is to calculate depreciation for nature as it is done for produced assets. In this context, *sustainable development* serves as a guiding principle. More precisely, this means – as a first approximation – making the use of material, energy, and areas for economic activities more efficient, but in the final analysis, sustainability requires the long-term preservation of nature's functions (potentials). Environmental-Economic Accounting is to show in statistical terms which natural resources are used, consumed, depleted, or destroyed by activities (production/consumption) of a period, and what expenditure is necessary for countermeasures; all this is based on the process of creating value added as reflected in economic statistics. Generally, only trends, mean values, distributions and similar macroeconomic indicators are of interest; individual cases such as materials, spaces, enterprises or incidents are aggregated.

Subject structure

The *origin of pressure on the environment*, the *state of the environment*, and environmental protection measures are the categories for which statistical data have to be provided. For pressures, a distinction can be made between material flows and area uses, and in the field of environmental protection there are aftercare and preventive measures. This subject structure is outlined in the flow chart entitled "Man – Environment – Man" (fig. 2.8.1).

Methodological concept

The calculation of depreciation for nature capital involves numerous methodological problems (problems of valuation/aggregation, limited knowledge of cause-effect relations, and great regional differences). For this reason, we must not expect too much of such a calculation. It would certainly be whishful thinking to believe that such a calculation could provide one single objective and indisputable depreciation value in DM terms from which a sound, sustainable growth of the national income could in turn be derived. It would be realistic instead to expect that in a gradual process of setting up such a system, the data actually measured, collected or

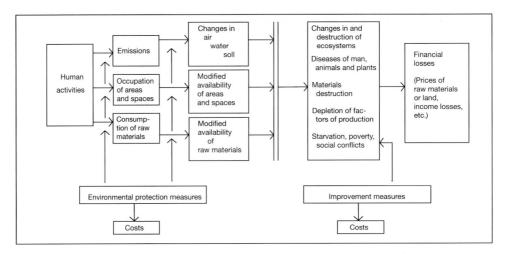

Fig. 2.8.1:
Flow Chart: Man – Environment – Man

observed would first be processed in a adequate manner, and then condensed further by means of standardized valuation procedures. It still remains to be seen to what extent such a condensation will be useful, and whether a valuation in money terms will be completely successful.

For this reason, Environmental-Economic Accounting has been set up in such a way as to provide answers to questions of economic and environmental policy at every stage on the was to the final system. For evaluating the efficiency of natural resource handling within the framework of structural and environmental policy, it is of fundamental importance to know how the use of raw materials, energy and land changes within the sectors of the economy over time, and what, in contrast, the emissions into the natural environment are. Highly aggregated indices of the state of the environment indicate qualitative changes in a standardized form and reflect the effects and benefits side of environmental protection activities which are actually being carried out. Prevention costs of additional preventive measures complete the picture, helping to weigh different "standards (target values) for individual serious pressure factors against each other and to decide in favour of one of them. Figure 2.8.2 presents the complete concept of Environmental-Economic Accounting:

As indicated by the different symbols, the various subject fields are characterized by their own specific methods: In subject fields 1, 4 and 5, methods of economic statistics and accounting are used to balance the material flows caused by the economic sectors and the environmental protection activities taken. Subject field 2 deals with immaterial pressures arising from a modified distribution of land uses, physical intervention, etc.; the methodological instruments used are remote sensing and geo-information systems. In subject field 3, the objective basically is to condense measuring and monitoring data, which are available in an isolated form, both with regard to space and contents, so as to provide suitable indicators; on the basis of subject field 2, an area sample is prepared which aims at the production of ecoindicators/ecoindices. The entire working area of Environmental-Economic Accounting does not include the setting of standards. For establishing such standards, however, information from Environmental-Economic Accounting explicitly aims at providing factual data, where available, on costs and benefits of alternative standard values for the process of political decision-making.

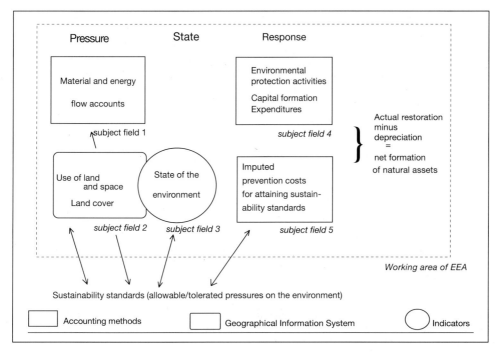

Fig. 2.8.2:
Environmental – Economic Accounting

Relation to national accounting

The result of the discussion on an environmental-related extension of national accounting is that it would seem best to continue as before with the traditional national product computations, which are an important means of short and medium-term monitoring of business development, and to create what is called a *satellite system*, a system for presenting economic-ecological relation separately, which has to be closely linked, however, with the traditional national accounts, its core system. This approach would be preferable in view of the methodological and statistical deficits still existing with regard to the valuation of economic pressures created by the business sector. Limiting the approach to supplementary satellite systems means that there will be the change to test new concepts and to use also data whose accuracy in statistical terms is not yet absolutely certain. This would not affect the data quality required for national product computations in the narrower sense. International concepts for a satellite system for the environment were developed in particular by the United Nations. In a handbook on national accounting, the System for Integrated Environmental and Economic Accounting (SEEA) was presented. In Germany, the satellite system for the environment is implemented on the basis of the conceptual proposals of the SEEA as part of Environmental-Economic Accounting.

Scientific advice

The Federal Minister for the Environment, Nature Conservation and Nuclear Safety established an advisory council for Environmental-Economic Accounting which has been entrusted with providing scientific advice for any issues related to Environmental-Economic Accounting. In 1991, the advisory council submitted an initial statement on the concept and the development requirements of the Federal Statistical Office, in which it stressed the necessity of reporting comprehensively and in a co-ordinated fashion on the links between the environment and economic activities. Irrespective of the still unresolved conceptual questions and the merging problems of implementation, the advisory council advocates that the work on this system be continued. It regards it as necessary to study partial fields in detail (and has also done so in the meantime), for instance the important area of valuation issues.

Work progress and first results

The subject fields provide the basic framework for further setting up and organizing development work. At present, research projects and field studies are carried out in each subject field with the assistance of external experts. Part of the projects have been completed; empirical data are available on economic activities creating pressures, emissions of individual sectors of economic activity and their expenditure for environmental protection, though still incomplete. In the future, project results should influence empirical activities, thus enabling a gradual approximation in statistical terms to the above change in "nature capital".

2.9 Standard reference data and policy assistance systems for global health evolution

C. Greiner, F. J. Radermacher and T. M. Fliedner

Introduction

In order to decisively support all types of initiatives for a sustainable development, that is following the general spirit of the Brundtland definition "*to ensure that it meets the needs of the present* without *compromising the ability of future generations to meet their own needs*" [WCED87], generally accepted standard reference data (as a minimum consensus) is of high value as a basis for projections and decision making. While it may be difficult to agree on the correctness, reliability, and adequacy of data in general, it is possible to characterize data as being either optimistic or pessimistic, e.g. bounding the real data in a reliable way (optimistic modeling approach). Such bounding of data allows correct inferences and predictions, as long as these are based on appropriate monotonic models and functions. To find consensus on such reliable bounding data within an international network activity, make it available worldwide, use it for all types of predictions, and monitor its correctness on-line is the focus of the research activity described, here. The chosen approach aims at a reliable description of the worldwide environmental and health situation, especially from the point of view of providing reasonable information, in the sense of a minimum consensus, as a basis for appropriate political decision making. This includes the demonstration of certain effects of measures taken today on the global future health and environment situation. It is taken into account that global health problems and the global state of the environment (including the overpopulation and poverty issues in developing countries) are closely coupled and, that it is rather difficult to make reliable projections in this context. A problem is that redirecting current trends towards a sustainable development is extremely difficult. However, there are many people, organizations, and networks in politics, religion, and grass roots communities all over the world working intensively towards this aim. It is the focus of the described research effort to support such groups with scientific instruments for the assessment and evaluation of various political plans.

The following text summarizes the state of the project activities of the Institute for Occupational and Social Medicine at the University of Ulm, in cooperation with the Research Institute for Applied Knowledge Processing (FAW), Ulm, and the Central Institute for Biomedical Engineering of the University of Ulm, in the above described context. As part of a two year pilot project, major conceptual and structural contributions have been achieved, which can now be used as a basis for a research activity that allows setting up a highly efficient, international, computer-based network of scientists who can contribute to reliable projections of future development on health and environment related questions on the basis of a common interface – the intended standard reference data.

Background

Physical and chemical systems can be investigated in experiments, which can be repeated if necessary. Learning about the complex system Earth is much more complicated. However, by running computer models based on different assumptions concerning strategies and systems dynamics (scenario analysis), we can get some insight into the behaviour of this complex system, and may identify reasonable paths into the future.

Much work has been done on studying the future of mankind and the living conditions of human beings on Earth via modeling techniques (see e.g. the models on behalf of the Club of Rome: WORLD3 ([Me72], [Me74]), the World Integrated Model WIM ([MePe74]) and the Bariloche-model ([He76]), further studies such as the UN World ([Le77]), SARUM ([Sa77]), Global 2000 ([Ba80]), Globus ([BrGr88]), and more recent approaches such as POLSTAR ([Ch92]) and various IIASA programs ([Fe91]) For a review of these models see [Br89]. Many such studies have forecasted alarming results. As all such models of the world and of human behaviour in this world are necessarily simplifications of reality, the predictions or of such of such studies are often relativized or even ridiculed. The Achilles' heel of most studies are the assumptions made as part of the modeling process and the adequacy and reliability of the data fed into these models. Thus the debate concerning the findings obtained is often centered more on the methodologies used than on the problem itself To increase the acceptance of models used to analyse complex systems is therefore crucial particular w.r.t. their use in practical politics.

Methodology

One way to increase the acceptance of global models concerning overpopulation and health related topics is to use optimistic modeling techniques to derive Standard Reference Data (See [Ra92]). This is an appropriate method because, according to the authors, it is not necessary to forecast precise data in this context, i.e. to concentrate on (overly optimistic bounds is sufficient. This means making the real situation look friendlier – the reality will be worse. Nevertheless, even these data and projections are rather alarming for many scenarios and different possible political actions. Therefore, even such optimistic data is often a reasonable basis for political decision making.

To achieve the aims described, it is planned to set up an international scientific network. In order to make international cooperation work, a clear interface and common denominator between participating groups is of critical importance. To identify this interface was one of the important findings of the pilot-project carried out in Ulm. In summary, concentration on a few key-variables should be used. For different regions of the world, only the following variables of data streams for the future will be – as a basis – considered in their dynamical evolution: number of people per region, internal production per region, and nettransfer between regions. Note that for example, the number of people per region may mean a lower or upper bound on the number of people living in a particular regions at a particular time.

Evaluation and validation of reference data takes into account available models, monitoring services, available prognostic instruments and various checks for proposed reference data against scenarios of future development studied elsewhere. In summary, through this interface many groups are able to work together quite independently, using and contributing to the data streams considered, with all of this eventually leading to a standard reference data stream concerning the variables considered. Whereby, all scientists involved will eventually agree that this reference data is correct in the bounding sense, i.e that there might be more but not less people per region, there might be less but not more consumable goods available, there might be less net-transfer from rich to poor regions but not more and so forth.

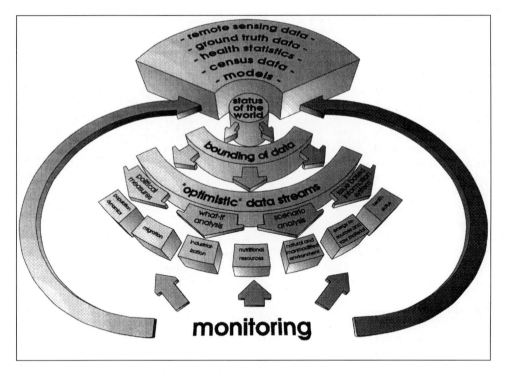

Fig. 2.9.1:
The concept for an *optimistic modeling* framework

On the basis of reference data, in the sense of future data streams of the interesting variables (number of people, consumable goods, net-transfer which are available through a consistent interface, predictions on all kinds of interesting health-related phenomena building on the reference data can be made. These includes, from an environmental perspective, the growth of mega cities, the shrinking of the rain forests, available water supply of a certain quality, available food, usable soil, and from a more health oriented perspective, the infant mortality rate, daily calorie supply per capita, average life expectancy, birth weight, level of education, access to health care facilities, and many others, always with reference to the basic data already discussed. If such submodels have certain monotonicity properties, standardized and reliable estimations in the sense of reliable bounds of the interesting information can be established.

Pilot study

The quality of models depends on the assumptions on which the model is built. Due to the reduced feedback and the many optimistic assumptions in this approach, not all effects will be obtained by simulations. But, as a reminder, this system is not designed for a precise description of the world. Simulations obtained with the chosen approach aim at an (overly) optimistic outcome. If even these predictions are rather frightening, then immediate action is required by governments and international organizations for a sustainable future.

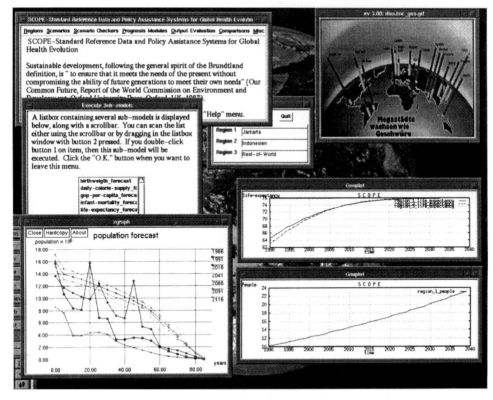

Fig. 2.9.2:
Screenshot of a session concerning a life expectancy and population growth forecast of Jarkarta

The investigations of the described concept has shown that a simplified and possibly unrealistic overoptimistic modeling technique can be a serious alternative or supplement to more detailed approaches in a number of situations [Fl92]. The chosen technique allows production of reliable bounds for data types which influence the well-being of humans, such as population growth, environmental development, economic development, life expectancy, education, and health, on the basis of an over-optimistic modeling technique.

The software architecture selected to support the chosen approach is characterized by its openness [Gr92]. It also offers the possibility of refining and coarsening elements of the framework without rebuilding the whole system (detailed modeling), and of integrating submodels (adding new constraints and coupling with other models) independent of the used modeling technique (Results of models like world3 and demographic models, such as [Ke68], [Ke77] and [Ro75], can be used to create input scenarios). Due to this architecture it is, in particular, possible to define a clear interface as a common denominator in order to make international cooperation work. Further tasks include developing manipulation options for scenario analysis, and testing and evaluating of methods for the analysis of different scenarios using also approaches from the field of artificial intelligence, knowledgebased systems, and fuzzy control.

A prototype that is able to demonstrate some of the main features listed above (creation and evaluation of input scenarios, coupling with other models, refining and coarsening of the system, reliable forecasts for selected variables, etc.) was developed at the FAW, Ulm. It is implemented in the language "C++", a superset of C which incorporates object oriented programming concepts. A user-interface in X-Windows has been developed, which provides a user-friendly environment, and which is especially useful for displaying results. The implementation is on a SUN Sparc2 workstation.

Summary

In summary, the feasibility of the chosen approach on the basis of an optimistic modeling technique has been shown, a prototype including the main features has been implemented, and guidelines were developed for setting up an international scientific network to make full use of this approach. Standard reference data and resulting predictions based on that data for special interest areas are a reliable data source, in the sense of correctly bounding behaviour of the future outcome. Proceeding in this way over the next decade, a reliable data stream concerning future developments will become available, and will constantly be made available to the general public. With corresponding prediction tools that can be organized in the sense of a "what-if" analysis, based on modern methods from the field of knowledge processing, it will be possible to support politics, ethic-oriented groups, and others. Such groups will then be better able to see the consequences of certain policies in terms of overoptimistic often still frightening – estimates. One would then have a defensible basis for public debate concerning resulting consequences due to certain policies. Such clear and generally available information will probably change the whole public debate on how to proceed and what to do. This might be a strong aid in reaching a more stable situation for our environment, an aim which is closely coupled with improving the health situation of humans on Earth from a global perspective.

Acknowledgments

The authors would like to thank Prof. Dr. Dr. J. Edrich for his continued advice and support and Dr. U. Bayer for her helpful discussions and suggestions.

This work is partially supported by the Senator Dr. Hans A. Merkle Foundation and the WHO-Collaborating Center for Global Modeling of Health Perspectives of the International Institute for Scientific Cooperation, Schloß Reisensburg, Gunzburg, Germany.

References

[Ba80] G. O. BARNEY et al.: "The Global 2000 Report to the President", Washington, U.S. Government Printing Office, 1980.
[Br89] S. A. BREMER,: Computer Modeling in Global and International Relations: "The State of the Art", Social Science Computer Review 7:4, 1989
[BrGr88] S. A. BREMER, W. L. GRUHN: "MICRO GLOBUS A computer model of longterm global Political and economic processes", Berlin, edition Sigma, 1988
[Ch92] Chadwick et al:: "The biosphere and humanity", in IIASA 1992 – An International Conference on the Challenges to Systems Analysis in the Nineties and Beyond, Laxenburg, Austria, May 12–13, 1992

[FlS2] T. DA. Fliedner et al: "Evolving Problems of Critical Significance to Global Health Interim Report of a Task Force of the Global Advisory Committee on Health Research", WHO, Advisory Committee on Health Research, Thirty-first session, Agenda item 8, Geneva, September 28 – October 2,1S92.

[FE91] K. Fedra: "A Computer-Based Approach to Environmental Impact Assessment", International Institute of Applied Systems Analysis, RR-91–13, Laxenburg, 1991.

[Gr92] C. Greiner: "Research Experience with and First Steps Towards an Optimistic Modeling Environnnent, in Modeling Global Development Processes and their Relevance to Human Health", WHO-Workshop, FAW-Ulm, Ulm, Germany, Sept.10–12, 1992.

[He76] A. O. Herrera et al.: "Catastrophe or New Society? A Latin American World Model", Ottawa, International Development Research Center, l976.

[Ke68] Keyfitz, Nathan: "Introduction to the Mathematics of Population" Addison-Wesley, Reading, Mass, 1968.

[Keg] Keyfitz, Nathan: "Applied Mathematical demography" John Wiley and Sons, New York, l977.

[Le77] W. Leontief et al.: "The Future of the World Economy: A United Nations Study" Oxford University Press, New York, 1977.

[N{e72] D. L. A Meadows et al. "The Limits to Growth", New York, Universe Books, 1972.

[N{e74] D. L. A Meadows et al.: "Dynamics of Growth in a Finite World", Cambridge, DJA, Wright-Allen Press, 1974.

[DAePe74] M. Mesarovic, E. Pestel: "Mankind at the turning Point", London, Hutchinson, 1974.

[Ra92] F. J. Radermacher: "Identifying Research Topics with Critical Significance for Global Trends Concerning Environment and Human Health", in "Modeling Global Development Processes and their Relevance to Human Health" WHO-Workshop, FAW-Ulm, Ulm, Germany, Sept. 10–12, 1992.

[Ro75] A. Rogers: "Introduction to Multiregional Demography", John Wiley and Sons, New York, 1975.

[Sa77] "Sarum 76 – Global Modelling Project", Report l9, U.K. Departments of Environment and Transport London, 1977.

[WCED87] World Commission on Environment and Development, "Our Common Future" Oxford University Press, New York, 1987.

2.10 Workshop results

2.10.1 Impressions at the close of the workshop

E. F. Roots

This morning, before the first reports from our Working Groups and Plenary Sessions were presented, Dr. Baumgardner asked if I would make some general comments at the end of our meeting, before we disbanded and went our separate directions. He suggested that It might be useful for someone who was not directly involved in the GTOS Planning Committee or HEM to give first impressions about what we seem to have accomplished or failed to accomplish in the past three days, and to say something about the steps we should be taking to move the data aspects of a "harmonized" Global Terrestrial Observing System from an idea into operational reality. At first I wondered if this request was a way of "making me pay" after Dr. Baumgardner had generously volunteered to be rapporteur for the discussions during the first plenary session that I chaired on Tuesday: – but this morning he has not even reported on that session! Joking aside, these last remarks at our Workshop provide an opportunity to reflect, at short notice, on what we have said to one another on this very important subject and to think a little, before the reports and final recommendations are written, about where ours discussions have taken us.

I will take as a starting point for these hasty comments the rather large questions that were opened and left open during the first session on "Data/Information Requirements". These, you will recall, had to do with lessons to be learned from existing regional and global activities; with our vision of GTOS 10, 25, 50 years from now; the requirements from the "user community"; new hardware; and research needs. All I can do, on the spur of the moment, is to share some personal impressions on how our discussions in the working groups and in the beer garden, and the points made in the admirably comprehensive and well-organized reports we have just heard, seem to relate to these questions and take us beyond them. I have tried to organize my thoughts, obviously without preparation, in a few categories – our vision; our "market", problems of support; conceptual and technical constraints; selection of priorities; areas we did not address or skipped over; and the actions we should take.

Need, vision and goal

The Fontainbleau and Geneva planning meetings that have given shape to the present concept of GTOS (including its relationships with GCOS and GOOS) have each made several statements, which are self-evident to the initiated, about why a coherent and comprehensive Global Observing System is necessary and justified. It has been noted that there are a variety of "users" who need information that is global in scope but on a variety of scales. As a GTOS planning group we have said, with fair confidence and no dissent, that the "policy users" need such information for national policy purposes, and for implementation of international and regional agreements and monitoring the effectiveness of those agreements. There have been statements that GTOS is needed so that government, institutions and the public can be kept aware, on an equitable basis, of changes and trends in environment, resources, and socio-economic geography (a new term that I learned this week) and that this awareness will help them make or change their plans or policies.

We have taken it for granted that the scientific community needs coordinated global observing systems to provide data that will advance knowledge of environmental and planetary processes, knowledge that will serve national and international policy, economic, and social ends. We noted that the sponsoring agencies for GTOS (FAO, ICSU, UNESCO, UNEP, and WMO) have each, in their respective ways, stated their reasons for support; and the Fontainbleau and Geneva reports contain statements and assumptions that information from GTOS will contribute to their activities and thus help them carry out their respective mandates. The UNEP-HEM, for its part, if I interpret the statements and comments correctly, has seen the Global Observing Systems and GTOS in particular as major activities that on the one hand will need the specialized services and coordination expertise of the Office for Harmonization of Environmental Measurement to bring about a successful observing system, and on the other hand be world-wide exercises that will help HEM accomplish its mission of enhancing the compatibility and quality of information on the state of the environment at all scales for all purposes. Thus the joining together of the GTOS Working Group for Data Management, Access and Harmonization, with the already planned HEM workshop on global harmonization of data was logical and fortunate.

These statements and the expressions of the need for a GTOS have been, I am sure, all justified and sincerely believed. But I felt that this workshop was not quite so sanguine as the previous planning meetings in assuming that just because a GTOS is needed, it will be wanted. There is a wide range of experienced people from many different agencies here at this workshop, and many of you have been quite clear in pointing out that in these days of conflicting policy priorities and fiscal constraint, not every good idea, no matter how logical or pertinent, gets supported; and that not all useful information is used in policy-making. Because of this uncomfortable but very real situation, it seems to this participant that it will for very important for GTOS to develop, as early as possible, a simple but strong statement of the vision, the need and purpose of a coordinated system for observing the global terrestrial environment. Such a statement must be in non-technical terms that are plain and convincing to policy people and economic decision-makers, and must make clear the usefulness, to those people at whatever level they may be, of the information that can be obtained only through a global observing system.

A "vision statement or flier" was called for at the Geneva planning meeting (GEMS Report 24), to publicize and explain the idea and scope of GTOS; but from subsequent developments and our discussions this week I feel that something much stronger and more convincing is needed. It seems clear that if GTOS is to be successful, it will need long-term national sponsorship from budgets other than the science and research budgets, in order not to be in competition with year-to-year research priorities. To secure this support, we need from the outset a strong and convincing non-scientific statement that can be used in all countries.

At the risk of saying more than I have a right to here, but I think reflecting what I take to be the opinion of several thoughtful people in this room, I might propose that it would be useful for GTOS, and the HEM involvement in it, to have an expressed long-term goal, beyond the policy and technical requirements, that could be seen by the supporting governments and their citizens as the reason behind the whole enterprise, and thus an activity worthy of their support. Such a statement might be in two parts. Both parts, it seems to me, have been expressed in different ways by the Working Group rapporteurs this morning.
One part would reflect the strong sentiment expressed in some Working Groups that the ultimate goal of GTOS should be to help create an informed public in both developing and developed countries which can, armed with this knowledge, demand improved environmental poli-

cies and actions from their governments. The other would address the need for harmonized and comprehensive environmental information with a global perspective that can provide a basis for improved scientific information and further research that in turn can be a basis for improved national and also local policies.

These statements might be phrased somewhat as follows (drafted as I have been listening this morning; stated here only to start discussion):

"The purpose of the Global Terrestrial Observing System is:

(i) to acquire and make available, in the shortest practicable time, coordinated and reliable information so that significant parts of the general public in all regions of the world can become aware of and informed or know that they can become aware and informed about the environmental characteristics of the planet on which they live, and be provided with information about how changes in the planetary environment affect their own environment and activities, and their options for the future. The availability of such information should encourage the public, national and international institutions to demand that policy-makers make and implement wise long-term environmentally sound as well as economically sensible short-term decisions.

(ii) to provide continuing and progressively updated information about the state of the global environment, its land, water, and living resources, on a full range of scales, to the scientific community and policy authorities, so that decision-makers at all levels will have the best possible scientific information on which to base considerations for alternatives for action and the likely consequences of decisions; and so that all decision-makers will know that others have accessibility to the same information and not be misled."

Clearly, much more thought and refinement must go into any such statement of goals. But the first important comment I want to make arising out of our workshops is that, transcending any single part of our agenda, GTOS should be acknowledged to have several clear purposes. Beyond the operational and scientific needs of its five sponsoring partners but including each of them and shared by all, the purposes of GTOS are to facilitate the availability of better environmental knowledge and awareness among the public in every country so that there will be an insistence on better long-term environmental policies; and to provide scientifically sound, coordinated and harmonized environmental information on which to base those improved policies at all levels.

The "Market": Who are the clients for GTOS and HEM?

Almost every statement that I can remember reading about the Global Observing Systems or other major data gathering or management systems – WCD's, CODATA, IGBP-DIS, BRIM, GRID, – has stressed that the data produced must meet the needs of the user, or be "user-oriented" or " be useful to the decision makers". We said the same thing, in several ways, this week. These statements are well-intentioned, but they often seem to be quite glib and perhaps a little naive if looked at closely. The blunt fact seems to be that those of us who are preoccupied with the business of measuring the environment or producing environmental data are not very good at putting ourselves in the position of the policy drafters or the so called decision-makers who as a rule want to be encumbered with as little information as possible, and who only want those bits of information that are seen to be directly relevant to the particular decision that has to be made at that time. Scientists, in general, are notoriously poor at communicating with policy analysts or political decision-makers in any form of government, in "policy" language. Given these generalities, and our expressed intention that our Global Observing Systems should be science-driven but user-oriented (Session 111; WG4), it would appear that, if GTOS is to have broad multi-national and multi-agency support and if the data it produces are to be fully effective, we have to put much more effort from our own side into learning who are our "users" and in assessing what information, packaged in what form, will be desired by those users as relevant to their work.

In the Working Groups there were many statements, rarely challenged by others, to the effect that the user we had in mind was the national government of our respective countries, or perhaps a UN agency; and at this level of generalization there seemed no point in being more specific. But what "market research" was under way or contemplated to determine who wants what global environmental data, for what purposes, and how much? It seems very important that GTOS does its "market research" before it begins to plan how to disseminate and package (i.e. "market") its product, especially when those who we think need the data are the ones who will have to pay, in advance, for the system that produces it. We must never give the appearance of having the attitude "We have an excellent product and a tool that can help you make better decisions:– you decision-makers should recognize its value and learn how to use this new tool." Experience shows that such an attitude is a quick way to kill a scientific or technical programme.

Some points in this regard that emerged clearly from our discussions were:
(i) there is no single or coherent decision-maker community. Those who can or in our opinion should make use of GTOS information include policy people at all levels from international to local, business and industries, scientific researchers, educators, environmental public interest groups, indigenous organizations, the general public and the media. Also, some countries and groups will not want additional information about the environment of their country to be freely available or used by others; we could call them anti-users. Within the rules of fairness and international relations, their interests and rights also must be considered.

(ii) different users have different concerns and priorities for global terrestrial environmental data. The priority subjects in the HEM proposal (Fig. 2.1.1.7 in Chapter 2.1.1), in MAB Digest No. 14 ("Towards GTOS"), and in the lists shown by Dr. Crain and Dr. Narjisse clearly show the difference in priorities and expectations between developing countries and industrialized countries. Within these various priorities, one also can segregate different kinds of data that are likely to be most important to environmental policy people, people concerned with political control or change, economic industry people, and environmental activists. Each or all of these "users" should have access to all data, if GTOS is to be truly a United Nations global observing system.

(iii) GTOS will not be a useful data system unless it is maintained in a consistent fashion over several decades. During that time, all user systems, and their inter-relationships, can be expected to change considerably. We do not know who the future users will be. GTOS must be designed now to meet tomorrow's needs by means of today's technologies, and be launched and supported by a decision system that evolved to meet yesterday's needs. However, its main value will be manifest a decade from now, in the next century. Its design and justification must be flexible enough to change as user needs change and technologies improve, yet it must be consistent and *harmonized* over time.

Support for GTOS

Several conclusions emerged clearly from our discussions:

(i) To be established and provided with continuing operational support, GTOS will have to be seen, by governments in several countries or by major international bodies, to be of short-term as well as long-term benefit to the government in power and to the present system of United Nations agencies.

(ii) Several of us felt that GTOS would succeed only if it had multiple sources of support, and – was not seen to be the responsibility of a single agency. But I think it is equally clear that it will not survive or even get properly launched, if it is an international orphan, a casual ward of half a dozen UN parents. GTOS, like the other Global Observing Systems, must have an identified champion and a principal host organization. The Memorandum of Understanding about to be completed between FAO, ICSU, UNESCO, UNEP, and WMO is an agreement to cooperate in design and planning of GTOS and to support the Planning Group (GEMS Report 24, Annex III), but does not assign responsibility. It is up to the GTOS Planning Group to address the question of responsibility and support. The various GTOS Planning Working Groups must address this together, in light of changes and financial realities in the UN System. I suggest that the report of this Munich workshop, together with HEM, should pay particular attention to the connections between the combined responsibility and commitment by various agencies to the operational aspects of GTOS on the one hand and the data harmonization and management responsibilities on the other.

(iii) The GTOS plan must be able to convince the present international data-generating data-gathering and data-management programmes (most of which are within the activities of the five sponsoring bodies, plus others in EC, OAS, OAU, etc.) that GTOS will be in their interest, not a competitor; and that it will increase their own effectiveness. These already established activities (there is a long list) are essential to GTOS, and the design plan as well as the arguments for increased effectiveness of their own data and information obtained must convince the on-going data systems to become willing partners and where appropriate to adjust their activities to become players in the system.

(iv) In order to justify support from national policy sources as well as scientific agencies, (See i above), GTOS must be seen by those who pay for it to be playing several roles. We have heard during this workshop that GTOS must be a *coordinator,* a harmonizer, and an integrator of data, even a *catalyser* to get observation systems under way. I think that we have had mostly the technical functions in mind when we discussed these roles. But to prove its worth to those who must support the system year after year, it is my feeling that GTOS will have also to give conscious attention to how it is going to serve as an *interpreter* of the significance of environmental data, as a disaggretor of environmental information from the general to local scale, and as a *stimulator* of demand, from the public and from policymakers, for better and more relevant environmental information. Some may feel that these are not proper roles for GTOS. But I suggest that the Planning Group should give this issue careful thought; for if these roles are not played within GTOS itself or by an international group close to it, the continued support for the Global Observing System may not last.

Problems to be addressed

As I listened to the Workshop reports and discussions this morning, I found myself making two lists in which I jotted down items that were stated as problems or constraints that GTOS will have to address. One group seemed to have to do mainly with clarifying and coming to a common agreement on what was included in the idea of GTOS. For lack of a better immediate term, I have called these "conceptual" problems. On the other list I noted problems that seemed to be mostly technical or scientific. Of course, there is no clear distinction between many of these; and some, I see as I look through my hasty notes, are re-expressions or corollaries of others. And both lists build on questions raised at the Geneva meeting, but become, I think, more specific as we get closer to "the real word". Here, for what they may be worth, are one person's quick summary of some problems that seem to underlie specific issues raised in the reports just presented:

Conceptual problems and constraints

– How to develop a coordinated environmental observing and socio-economic data gathering system that is global in scope, science-driven but politically supported, and yet have it respect national sensitivities and priorities? This is perhaps the most difficult question of all, for on it hangs the support, the national and international and scientific credibility of GTOS and the other Global Observing Systems. The way that this problem is addressed or solved will have much to do with whether GTOS can be initiated as a global system, and how long it will last. The Fontainbleau and Geneva planning meetings recognized this problem, but very properly, at that stage, concentrated on what was the GTOS that the planners had in mind and how should it develop in a technical sense before worrying too much about how it would become established in the world of international relations, finance, national priorities, cultural sensitivities and scientific exclusiveness. But now those issues will have to be faced Here is where GTOS will need United Nations sponsorship and experience.

There are a number of sub-issues that arise from this question, that were aired in our Working Groups and their reports. Let me list the ones I noted in the order I wrote them down, not in order of importance:

(i) do we need a pre-meditated and transparent way to deal with cultural suspicions about making freely available to everyone, information about one's own and everyone else's environment? Or would it be better to deal with such issues, when they arise, quietly and unobtrusively according to the culture concerned (and thus, perhaps, not through the nationalgovernments)?

(ii) will it be necessary to set up in advance some method of preventing unwanted commercial or political exploitation of environmental data, to the detriment of some countries or some sectors of society, as for example in the issues raised by Dr. Pykh?

(iii) how will GTOS ensure that it maintains accessibility to data and observations gathered and held by other observing systems which will be valuable to GTOS but whose support and mandate will be from other sources for other purposes?

(iv) how will international and scientific credibility for GTOS be earned? Scientific credibility will be little problem if the data and information, whatever their quality and completeness (and these are bound to range widely) are well documented and transparent as to their criteria and standards, methodology and replicability. But what about political and social credibility? We know that observation of some environmental characteristics, such as surface temperature or sea level, can achieve credibility on purely objective technical grounds; but observation of other important parameters, such as changes in forest cover or soil water content, involve some subjective elements that we know from experience bring the data to the edge of policy and "credibility" involves more than objective measurement.

(v) how can GTOS demonstrate the "public good"? Such demonstration will be needed if GTOS is to be supported, over the years, as a continuing national responsibility in many countries, and not according to one-year-at-a-time science priorities and budgets.

(vi) how will GTOS be structured so that everyone interested (and every country) can take part, and yet have consistency and firm priorities so that the most important things are observed and efforts and resources are not dissipated on "unimportant" observations? How "democratic" or how "dictatorial" should GTOS management be? Will "control by consensus of active players", as in the Environmental Protocol to the Antarctic Treaty, be workable for a Global Observing System? Or should there be an international "dictatorial" authority, as in WCRP?

(vii) who will develop, and how, the means for feedback of GTOS data in the form of easily accessible, understandable information to identified interest groups or user groups? How will this be paid for? If GTOS itself is to act as an interpreter and disaggregator of its own data (see 3 above), how can it do this without being accused of drawing premature conclusions or making predictions? These kinds of problems lead also to the need to make clear the involvement or otherwise of GTOS in explaining the observed data and environmental trends in non-scientific terms, and in local languages where this may be desired, in many countries around the world.

(viii) what should GTOS be doing to cultivate public and policy demand for better environmental information on a global as well as local scale, and for creating the capacity among "users" to make the best use of GTOS information when it is available? Should GTOS be developing a manual, a "Users' Guide to GTOS"?

A further "conceptual" issue, and one that is also largely technical in nature, which has surfaced at this Workshop is the problem of coordinating the development and support of GTOS with that of GCOS and GOOS. The three Global Observing Systems are taking somewhat different approaches, but together they must mesh to fill a world-wide need. We must take care that the three Global Observing Systems are mutually supportive not in competition with one another.

A useful and thought-provoking way at looking at how the science aspects of GTOS relate to the "real" world of money, policies, management and cultural issues was shown this morning to several of the workshop participants during the brief presentation of the German Terrestrial Ecosystem Research Network (Deutsches Netzwerk Ökosystemsforschung). Dr. Lenz has kindly agreed to let me use his figure showing how the world of science relates to the worlds of planning, engineering, technology, experience and culture (Figure 2.10.1.1). I would ask each of you to consider carefully some of the relationships shown, in light of what we think GTOS should be and should do. GTOS, science-driven but politically supported and with policy, planning, and cultural clients, must overlap and in some ways bring together these worlds.

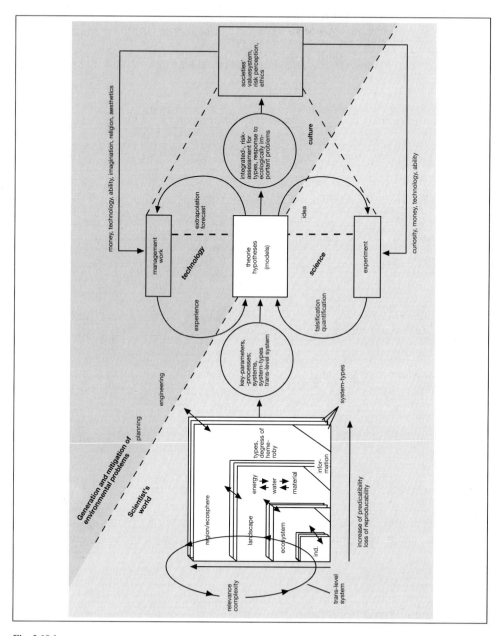

Fig. 2.10.1:
"Environmental Systems and Theories in a Transdisciplinary Context" – by kind permission of R. J. M. LENZ, GSF – Research Centre on Environmental and Health; Munich, Germany.

Technical problems and constraints

The technical and scientific problems to be faced by GTOS are in many ways the main content of our Working Group reports, for most of the questions we were asked to consider were designed to identify and discuss the desired scientific content and technical operations of a global terrestrial observing system. My list, therefore, is hereby restricted to only four headings, which will not be elaborated in these comments but which must have adequate technical discussion as GTOS develops. These problems include:

(i) *programme design* – e.g. how to relate observations by different technologies of the same phenomena, as for example remote sensing and on-site information;

(ii) *scaling problems*, in space and time. This issue partly a matter of programme design, partly a "conceptual" issue (see iv(a)), and partly a question of knowing what the users will find most useful (see ii). But it is also a technical matter, and how GTOS handles the choice of scales and changes of scales will perhaps more than any other characteristic determine the usefulness and thus the long-term viability of GTOS.

(iii) *choice of measurements, sensors, recording methods*. These of course determine "what GTOS is". They are a matter of programme design, scaling, vision, etc.; but they also must reflect available and evolving technologies, economics and world-wide practicality, and data management and transmission systems. An important consideration is to ensure coordination, where appropriate, with GCOS and GOOS and other world-wide systems such as those of WMO and the geophysical networks. A much more difficult problem, yet to be addressed, has to do with the most useful parameters to be observed, and how to observe them, to obtain compatible georeferenced socio-economic data as they affect the environment in all parts of the world.

iv) *information delivery and information technologies*. At the sophisticated technical level, this is a rapidly developing and fast-changing field, and GTOS will need to keep up-to-date as it continues through the next decades. But as a U.N. system, GTOS will have to ensure that its information is accessible to, and useful to, every important user, whether technically sophisticated or not. Therefore, GTOS data must also be delivered in simple easily understandable form. And the data we obtain today must be in a form that can be compared and analysed with what will be current data fifty years hence.

All of these problems, and others, imply that GTOS must be supported by a programme of continual research. Many of the research priorities have been identified in the individual Working Group reports.

Selection of priorities

It was clear to all of us that a Global Observing System cannot attempt, and should not pretend, to observe everything on planet Earth. And it must be careful not to raise expectations that just because it is a global system, people can get from it any information they want about any part of the globe. But those who support or pay for a GTOS have reason to expect that it will be able to deliver information of the kind they think will be useful, on a global basis. Therefore the selection of priorities for what should be observed, how often, to what standards using what techniques, will be a key factor in the credibility and usefulness and thus the viability of the whole system. It will also be essential that the priorities for GTOS observations be clear to all, but not rigid.

Several participants pointed out that the priorities will change with time or with events – that what is most important to observe today will likely not be most important a few years hence. This will inescapably be true as GTOS learns how to obtain and make use of socio-economic environmental data. And changes in sensor technology or advances in understanding of environmental processes could cause important changes in priorities or targets of observation.

It was the feeling of some that GTOS should not try to set its priorities all at once and emerge full-blown, as it were, from the planning discussions. Instead, some thought GTOS should start modestly, build on some compatible observing networks already in place and which were at present obtaining environmental information which is of high priority to the present sponsors, and expand from there, learning as one goes and if necessary changing priorities with experience and changing demands. Many of the scattered notes I made this morning on the subject of priorities are, I now see, along this line:

(i) GTOS priorities should be user-driven. This statement may contradict what I reported earlier that the Workshop felt that GTOS should be science-driven but user-oriented; but I think what we had in mind here is that the most important things for GTOS to observe are those features of the environment that can provide information of value to policy-makers and the public, and not primarily those features which it is a scientific challenge or fun to measure. There may be some justification in the criticism that some satellite observations gather enormous amounts data of impressive accuracy on characteristics of Earth that no one cares about and not one can figure a use for, just because some scientifically exciting sensors have been put into orbit (as a one time member of a NASA satellite programme sensor selection team I wince at this accusation, but cannot completely deny it). We do not want GTOS to be subject to that kind of criticism.

(ii) GTOS should be global in scope, but its priorities may be any scale, from local to global.

(iii) The priorities and criteria for GTOS observations should be shaped not only with the users in mind but also with the expert help of those who must handle and massage the data. What is the best way to measure the most important characteristics of the environment may well be influenced by the way the data are handled and how they can be harmonized, correlated, or synthesized. Thus GTOS planners should work with statisticians and HEM experts in setting priority targets and designing the observing programme.

(iv) We agreed that a first step in setting GTOS priorities is to select the key parameters to be measured or observed. One of the general objectives agreed upon at Fontainbleau was to select the minimum set of processes and variables needed to detect global change and validate models (MAB Digest 14). The Working Group reports noted three areas that should be given attention in the selection of priority parameters to meet this objective:

(a) the lists of problems, gathered from policy-makers and other users, for which there has been an expressed need for global data or for which the Planning Group, in its wisdom, feels that global data will be useful. Such lists are an obvious staring point. But their inconsistencies or incompleteness places an additional but very important onus on the Planning Group, that of identifying the gaps where there are no data and also identifying gaps between problems, which could be identified and addressed if there were adequate data. These may be gaps that have not been noticed, or, more likely, gaps because no one has yet been able to et funding to address.

(b) the need for core data relevant to in the various on-going activities of the sponsoring agencies – WMO, ICSU, UNESCO, UNEP, FAO – already determines priority observations that, at least in part, should be obtained ore effectively and economically by a GTOS than through independent observing systems. At the same time, several of those programmes are already producing data valuable to GTOS and can continue to do so within their on-going mandate – WMO has obvious examples. Part of the job of planning and designing an effective global observing system will be to select, from the presently on-going activities, those elements of priority for GTOS.

(c) the priorities for GTOS should take into account their likely contribution to GCOS and GOOS. As each of the main elements of the Global Observing System is being organized in a different fashion and is in different stages of realization, it is incumbent on the GTOS planners to keep good liaison with the climate and the ocean system planners, so that a sense of shared priorities will emerge.

(v) The selection of main sites for GTOS ground observations and in situ research is an essential part of priority-setting and programme design. Workshop participants pointed out the need for establishing some criteria for GTOS sites and the importance of reviewing in a systematic way the obvious already established candidates: UNESCO/MAB Biosphere Reserves, ITEX sites, ICSU/IGBP sites for START/RRC, GCTE and LOICZ, etc.; national and regional networks such as LTER, TERN, etc. Not all of these will be suitable but some will likely be excellent, as demonstrated on the posters at this Workshop, and they can get GTOS off to a good start in many parts of the world. Use of these sites will also have the advantage of bringing other scientific programmes into the Global Observing Network fold as participants. Participants noted that GTOS would be wise to develop two general cate-gories of sites: – a relatively small number of sites, in representativity major terrestrial eco-regions (could we use the 14-region START classification to begin with?), and a larger number of intermediate or comparative sites that will add data and help establish typical representation. These issues had also been discussed at Fontainbleau and Geneva, where it was pointed out that, perhaps in distinction to the IGBP sites, the GTOS should include intensively managed landscapes, urban areas, and highly industrialized locations (and the socioeconomic data from them) to provide information about their role in the world environment.

Attention was also drawn to the need for GTOS to make use, in the beginning particularly, of "sites of opportunity" or perhaps "sub-regions of opportunity" where there are already small-scale observing systems under way that can serve both as models and starting points for GTOS. The United Kingdom Environmental Change Network studies, or the German MAB program on integrated monitoring and research in the Bornhoved Lakes District were seen as possible examples. The MAB BRIM (Biosphere Reserves Integrated Monitoring) compilation of sites and research activities, the MAB Northern Sciences Network roster of circumpolar research sites, and the information on protected areas of the world available from the World Conservation Monitoring Centre were seen as very useful sources of information to help in selecting a priority network for GTOS observing sites None of these, however, appear to be sources of data on representative socioeconomic studies.

Areas not addressed or skipped over

Naturally, in three days and with a structured agenda that kept us busy on very worthwhile but focused lines, we could not discuss or even think of all aspects that will be important to observing, on a systematic basis, the environments of all continents of the world. There were some subject areas that came up in our discussion that seemed to be important but which we did not pursue because they were off the subject of the moment or because we did not have time. The following is one person's quick list, not yet even reviewed for the first time, of subjects that seem to me to be pertinent to a GTOS as it seems to be evolving, but which we did not address here at Munich.

Social, socioeconomic and industrial observations and data

Although the GTOS framework agreed to at Geneva stated that its focus will include "the human populations and settlements and the results of their impact on the land", and there were several statements here in Munich that GTOS must include social and socioeconomic data, we did not discuss the "human" side of a global observing system except in the context of political, economic or cultural constraints on access to data. What we missed and what must have more attention before GTOS is launched, is a plan of how social, socioeconomic and industrial data, needed for GTOS, are to be obtained; how criteria and standards for them are to be established on a global basis; and how they can be harmonized and amalgamated with biophysical data. It seems to me that the Planning Group might consider the role that environmental indicators, perhaps along the lines of the "environmental performance indicators" used in the Netherlands, or possibly in the paper distributed by Dr. Pykh (which I confess I have not yet read) might have in selecting means for obtaining relevant social and economic data; or perhaps the "industry as an ecosystem" approach developed by William Clark for the US Global Change Programme, may be useful. Or the modeling of the "ecosystem cultural gradient" in the German FAM project may point in a promising direction. In any case, this aspect of GTOS needs early and serious attention.

Relationship between remote sensing and ground-based studies.

This aspect of global observing is a key element of program design, sensor and site selection, and scale management. Several times our discussions touched on the need to incorporate all kinds of observations and "harmonize" them, but we did not address them. Henry Nix gave us some good examples from, for example, Tasmania; but the lesson I got from that example, confirming my own experience in the arctic, was that bringing together a lot of different kinds of observations will not happen without much hard work by knowledgeable people who are convinced of the advantages of doing so. The same kind of issue, on a different scale, will certainly come up when GTOS, according to its own framework plan, starts to include microbiota and their behaviour. Only in this case it will be amalgamation of micro-remote-research data and macro observations. I think that GTOS will need some specialized expert workshops on this subject, perhaps with input from GCOS and GOOS and IGBP core programmes.

Averages or hot spots?

Another area of programme design and scaling that we touched on but skipped over is that of whether, and how, GTOS should attempt to show representativeness, and broad global or regional trends, or give attention to anomalies. Should it focus its main efforts on observing impacts and present changes, or be designed to provide early warning of impending changes? The report from the Geneva meeting says that GTOS should do all of these things; but obviously it cannot do them all equally. Should the system be structured to be as representative of national ecoregions and collective human influences as possible, thus avoiding anomalous and known sensitive areas; or should it be designed to keep an eye on danger zones or potential threats to the environment? Should GTOS, for example, be a source for documenting the spread of zebra mussels which are at present mainly important to only one continent, or should it limit its priorities to issues of global or multi-continent concern? Such questions have relevance not only for the scientific validity and "balance" of GTOS, but also for the likely interest and support of the governments and the public of the countries which must pay for the programmes. They also have relevance for the scientific groups and monitoring activities that will either participate in, supplement, or operate independently from GTOS. These issues should be carefully discussed early in the planning of GTOS, and clear decisions made; which should be reflected in the programme design and the "vision statement".

Relation to other international data systems

This workshop has been an expanded meeting of the Data Management, Access and Harmonization Working Group of the GTOS Planning Group. And several of the participants here are persons currently or recently involved in other international scientific data systems connections with WDC's, CODATA, BRIM, GRID-IDDEA, IGOSS are here in this room. It seems strange, now that we have come to the end of the workshop, that we never gave any real attention to how the GTOS data or data management will relate to or make use of the experiences, good and not so good, of these other systems. All data and data management systems in the world today are having a hard time because of financial constraints and reluctance of many governments to support anything that is long-term in nature or which has interests beyond domestic boundaries. The data management systems that will be the "product" of GTOS will have to be compatible with, or support, or be supported by, what is presently in place, and the differences and relationships will have to clear to outsiders as well as insiders. This is another area that needs early GTOS attention.

Reference materials and samples

An important part of any long-term observing system is the preservation and archiving of critical materials and samples. The principles for design of GTOS slated at Fontainbleau include the need for archiving samples as well as data for future analysis; and this was also a stated necessity for long-term harmonized environmental measurement emphasized at the time of the establishment of HEM Working Group 5 noted the need for a strategy in this area. Let us not forget it as GTOS evolves.

Procedures for "harmonizing"

Although UNEP-HEM is a co-sponsor and host for this workshop, the attention has been, by intention, on GTOS and its data requirements. But the GTOS Working Group on Data also has "Harmonization" in its title. It does seem, in short-term retrospect, that an opportunity was missed, by both HEM and GTOS and by us all, to have a good discussion on what is needed and is meant, in the 1995-2000 context, by "harmonization of environmental data". What are the practical elements and specific goals of harmonization? What are the procedures? What are the criteria, and who sets them; and how does one know when one has done it? As someone who, as I explained last Tuesday, has been making noises about this subject for a while, I confess I am still not very clear what it all means; but it did seem that this workshop, using GTOS as a vehicle, was a good place to explore the question in a modern and practical context. Perhaps it has in fact been so; but we did not openly discuss it as a subject in any of our sessions in Working Groups. I do hope that out of all our discussion comes a positive message to Dr. Keune and his group on the issues and elements of harmonization of data with respect to an operational international programme; and that the GTOS Planning Group and the rest of us have a new appreciation of how necessary and valuable an HEM Office is to our own activities.

GTOS as a research stimulator

One important subject area, which I think was in the minds of the sponsors of the Workshop and several of us when this event was first discussed, but which was not openly mentioned in our sessions, is how GTOS, or the family of Global Observing Systems of which it is a member, can serve to stimulate and support the research by national institutes to improve the scientific strength and quality of the observing systems themselves. Despite the nice words about the need for integrated monitoring from UN agencies for many years, and from Agenda 21, despite the positive expressions of interest and need from each of the five U.N. sponsoring agencies for GTOS, and despite the conclusions of every Workshop Group here at Munich that continuing research is needed if GTOS is to be an adequate global observing system over the next several decades, the blunt truth is that in nearly every country here represented the funds for research on environmental monitoring are under threat or are steadily being reduced. And funds for research to develop ways to gather social and economic data in a georeferenced environmental context seem to be essentially non-existent. We did talk a little about this on the first evening in connection with studies of the FAM project in Bavaria, and some of us cried in our beer over it at the Biergarten; but as far as I know it never was addressed in the Workshop. The research needed to improve the science in GTOS will have to be undertaken by national - research dies, for essentially national purposes which will have international benefit. Let us hope hat the recommendations to the GTOS Planning Group and the report from this Workshop ill give us ammunition – and resolve – to make a strong case, each in our own countries, o re-invigorate research on environmental monitoring and measurement.

Actions: What next?

Each of the Working Group reports that have been presented this morning makes some recommendations on actions that should be taken to put GTOS into operation and ensure that its data are properly managed and harmonized. Several have identified areas where research is needed. It would be presumptuous to comment on these without studying them carefully. But because Dr. Baumgardner asked me to try at short notice to give some overall impressions of where we have got to in this Workshop and comment on where we should go from here, the following seem to me to be some of the most important things that now should be done, in rough sequence:

1. Decide more firmly on the objectives and scope of GTOS, building on the framework that is in the report of the Geneva Meeting (GEMS 24) and the ideas of the Working Groups of the Planning Group; coordinate where appropriate with the objectives of GCOS and GOOS; and prepare a careful objectives and vision statement in non-technical language that will convince governments and international agencies of the necessity for and value of a long-term GTOS.

2. Develop a transparent and scientifically comprehensive way to select the priority parameters for GTOS observations, and make the method of selection known to the scientific community, to potential user groups, and to other international data-gathering programmes, so that a wide range of those who need the data and those who know how to make the observations can have input into the ranking and the choosing of parameters.

3. Put energy and planning into making contacts with those involved in social and economic data relevant to environment and national resources; determine the present availability of social data relevant to GTOS and devise approaches to improve that data, fill gaps, and re-format the socioeconomic information to make it compatible and harmonized(?) with biogeophysical data; do "market research" to identify the potential users for socio-environmental data, and determine the format and mode in which such data will be most useful to them.

4. Contact future partners in GTOS, including the obvious global ones in ICSU, MAB, IHP and WCRP programmes as well as GCOS and GOOS, and the programme operational people in the UNEP family (GEMS, GRID, MARC WCMC, HEM) to ensure that GTOS will be as compatible and cooperative as possible with other activities in the general field. The same should be done with regional programmes such as EEA, and specialized groups such as space agencies who are gathering data, or have responsibility for improving the capacity to make observations and gather data, on the terrestrial environment and resources. On the data management and dissemination side, contact or establish liaison with the World Data Centres, the U.N. Statistical Office, and others who will handlers and purveyors of GTOS data. Many or most of these agencies are of course already in some way involved in the planning of GTOS and some of their representatives are here in Munich; but what is needed at an early stage is deliberate broad contacts at the appropriate administrative and support levels to ensure that GTOS does not appear as a surprise competitor to already established and worthwhile activities, and to develop the partnerships that will be essential if GTOS is to succeed.

5. Only after progress has been made in each of the above areas, does it seem sensible to begin seriously to design the GTOS programme in terms of sensor selection, vehicles and modes of operation, sites for intensive observation, intermediate observation location or transects. etc.. and data management schemes. Naturally, some very preliminary ideas on what to look at and how to do it must be in laid out and exchanged at the beginning, as was done at Fontainbleau, as a basis for establishing a convincing need for GTOS and for developing its objectives. But I think our experience with IHP, GARP, and IGBP show that scientists are apt to rush too quickly into programme design before they have fully explored their objectives and scope, or identified their clients and partners. Let's not make that mistake with GTOS.

It is at this stage, particularly after partner programmes and data handling facilities have been tentatively identified, that realistic cost estimates for GTOS and its components can be made. At the same time, some plausible hard-nosed benefits to major users should be identified in economic, policy, or international relations terms

6. Coincident with or closely following on the preliminary programme design for GTOS must be development of a "science plan" that will link GTOS to the on-going scientific and research activities of participating countries and international science programmes. In a normal scientific or research proposal, the science plan comes first, and usually incorporates pro-

gramme design. But the Global Observing Programs, as they are developing, are not primarily scientific activities, although they should be based on the best science we can bring to them and they should contribute in an important way to future environmental and global science. They are programs for which science is essential, but their purpose is to provide data and information to decision-makers and a wide range of non-scientific users. The "science plan" therefore, should be subsidiary; it should be directed toward ensuring that the scientific basis for the observations is sound, that the observing methods and technologies are the best or are adequate, and that the measurements obtained are indeed what they purport to be. It should also ensure that the interpretation of the measurements or observations in terms of its environmental or resource significance has a sound scientific basis; that the handling of the data and its harmonization from different observing methods or on different scales makes scientific sense.

Equally important, perhaps, the scientific plan should give attention to how GTOS procedures and their data contribute to on-going research programmes in the many areas that will be involved; and to how the future advances in scientific understanding of environmental processes or in observing technologies can be incorporated into GTOS as it evolves.

We must ensure that the science plan includes plans for establishment of data-bases on a range of scales (a very difficult scientific problem), for data management and responsibility, internationally and over the long term; for archiving of data and specimens; and for development of theoretical concepts and technical skills in the use of regional or global data in assessment of environmental change.

Possibly the most difficult area for development of a science plan for GTOS, as seen from where we are at this stage, will be how to incorporate and mesh the scientific methodologies and priorities of the human and social sciences with those of the natural sciences into a coherent system of observations of the natural world, of the humans within or on it, and what humans and their works are doing to the environment and resources. Both the social science community and the natural science community are engaged in research in these fields, and both need and use data on a range of time and space scales; but their approaches to obtaining and using such data seem to have been independent and different. GTOS - requires that they get together – harmonize, Dr. Keune would say –, and while this will not be easy, in the long run it will be good for science.

7. On the basis of all the above, the GTOS planners (or will they be called something else by this time?) will have to concentrate on obtaining support and long-term commitment. The support must be financial, in-kind from cooperating and partner agencies and programmes, political or policy-related at several levels, and from the scientific community at large. I have the feeling that unless there is real, long-term support from each and all of these four areas, GTOS and its sister Global Observing Systems will not get very far. There is good support, and even enthusiasm from some of these quarters, for the initial planning, and that is what has brought such a useful variety of people here to Munich. But support for planning is one thing, and support for keeping and operation going over the long haul, in good times and times not so good, is quite another. As we have said several times in the last couple of days we should be planning an observing system that will be most useful if it lasts at least 50 years. For this there must of course be long-term financial support. But equally important will be sustained policy interest and involvement of the changing scientific community . The only way to achieve this will be to ensure that GTOS observations and data are useful, and before long are taken to be indispensable, to each of the groups from which

support is required. The objectives, vision statement, partnerships and science plan of GTOS are vital to developing a broad conviction of need for the data, which is essential to the realization of policy and financial support.

8. Notwithstanding all of the above but including them, it is important that GTOS get started and into operation. One must take into account all of the problems and complexities and diverse factors, but it is necessary to get or with the job. One could plan for ever; a global scale programme can never be fully planned in advance. There are a number of activities now under way that can logically become part of an operational GTOS; some should be identified, labeled as initial actions of the Global Terrestrial Observing System, and an announcement made that GTOS is in business. Without in any way decreasing the importance of items 1 to 7 above or the recommendations made by the various Working Groups, one of the best ways to have an effective GTOS is to get it started and "learn by doing". It is also easier, as a rule, to get support for an activity that is actually in operation than for an idea still being argued about. So let us get on with it.

2.10.2 Summary report of the workshop

M. F. Baumgardner

Background

An important event which preceeded th Workshop on Harmonization of Environmental Data but which had an important bearing on the objectives and implementation of the workshop was the first meeting of the Ad Hoc Scientific and Technical Planning Group (STPG) for GTOS (Global Terrestrial Observing System) held in Geneva, Switzerland, 13–17 December 1993. The major accomplishment of the Geneva meeting was the naming of three specific Working Groups and a chair for each WG, and the development of Terms of Reference for each WG, including an 18-months schedule for implementation of these terms of reference. Each of the three working groups will contribute to the development of a final report of recommendations related to the implementation of GTOS to be submitted by the STPG to the international sponsors, four agencies of the United Nations (FAO, UNESCO, UNEP, WMO) and the International Council of Scientific Unions (ICSU).

Working Group 1 (WG1) has responsibility to report to the STPG by mid- 1995 its recommendations for strategies and a plan of action for "data management, access and harmonization" for GTOS. The chair of GTOS WG1 worked closely with the Director of UNEP-HEM in planning the agenda and strategy which was followed in the Munich Workshop. From the point of view of the members of GTOS WG1, the Munich Workshop was the initial working meeting of the Working Group in the implementation of its terms of reference.

A group of thirty international scientists and representatives of international sponsors assembled in Munich, Germany, from 7 to 10 June 1994 to participate in a Workshop on Harmonization of Environmental Data. Hosted and organized by the UNEP-HEM (Harmonization of Environmental Measurements) office in Munich with the support of the German Federal Ministry for Research and Technology, the Workshop focused on issues related to harmonization of environmental data, especially at the global and regional scales, in support of the proposed Global Terrestrial Observing System (GTOS). This Workshop is one of a series of meetings which will contribute to the definition and planning of GTOS.

Perspective

The Workshop was designed to build on previous studies of monitoring and assessment of terrestrial ecosystems. The Workshop focused on the management, access and harmonization of environmental data for terrestrial ecosystems from the time of data acquisition to the delivery and use of environmental information by decision-makers and policy-makers.

The agenda for the three and a half day meeting was developed as a workshop. On the first day a few invited presentations set the tone and scope of issues critical to the development of GTOS for consideration in the deliberations which were to follow. Six working groups of less than ten persons per working group were asked to address a specific set of questions related to the focus of the Workshop. The working groups were asked to consider the problems related to the removal of a wide range of constraints to management, access and harmonization of environmental data for a Global Terrestrial Observing System. These deliberations were intended to provide concrete guidance for the further deliberations of GTOS WG 1 of the STPG for GTOS. The main points discussed and agreed in the working groups and plenary sessions have been collated and summarised under three specific areas--set of principles which should establish the nature, structure and operating guidelines of GTOS, a definition of GTOS and its objectives for rationalizing and justifying activities of an operational Global Terrestrial Observing System, and a set of recommendations for research priorities, specifically focused on GTOS data management, access and harmonization, in support of GTOS and related Earth observing systems.

Set of principles

Operating principle of GTOS

GTOS will be designed to operate on the principle of data integration within a systems paradigm, where the links between data items are explicit. If the system is to be predictive as well as documentary, it must provide the information needed to assign a scientifically credible cause to the observed effects. It must ensure that the driving variable, the state variable and the response are all collected and analyzed at compatible and comparable space and time scales and accuracy.

GTOS must operate on the principle that a major reason for being is to document change and to attribute causality by strong induction.

This imposes considerations of sampling adequacy, quality control, and linkage of variables with an a priori model which predicts a specific observable change.

GTOS data management

Data management activities will be subject to a routine external review.

Before entry into the GTOS database, all imported data will be screened, assessed and cleared for entry.

Data for GTOS

A basic principle for GTOS will be the assembly and distribution of data/information which focuses primarily on global environmental issues.

GTOS will operate under the concept of the free flow of data/information always in accordance with the principles of reciprocity and equity.

Those who provide data to GTOS shall have access to the GTOS network.

GTOS will follow the principle of retaining primary data at the highest level possible.

GTOS will follow the principle that all GTOS data must be documented by an appropriate (minimum set of) metadata.

A basic principle of GTOS will be the harmonization of data collection, exchange and assessment of environmental data (biotic, abiotic and socioeconomic).

Distribution of GTOS data/information

GTOS will make use of existing networks and global, regional and national communications systems for the distribution of GTOS data/information.

GTOS data quality

GTOS will support the principle of highest scientific standards (objectivity, reliability, validity, representativity) (of Chapters 2.2 and 3).

GTOS will support a plan of quality assurance as a way of sharing knowledge, experience and capability to enable all network users to achieve the same high standard.

Intellectual property rights, privacy acts, copyright laws
GTOS will follow the principle of acknowledgement of data sources and rights of those who generated and/or assembled the data.

Technology transfer

GTOS will support the principle of capacity-building to strengthen the capabilities in all participating nations to have access to and utilization of GTOS data. A strategy which has been very effective in capacity-building in some situations has been that of "twinning" of institutions. That is, an institution with excellent existing facilities and experienced scientists may be "twinned" or matched with an institution in a country which is desirous of developing the capacity (hardware, software, human resources) to perform the environmental monitoring and research required by GTOS and so important to one's own country and its development. GTOS might play a catalytic role in arranging such twinning.

International collaboration

GTOS will operate on the principle of open and intentional cooperation and data exchange with GCOS, GOOS and other international environmental observing systems.

GTOS will operate on the principle of open and intentional cooperation and data exchange with national environmental and related agencies and other organizations and institutions which make terrestrial ecosystem observations and measurements.

GTOS will uphold the principle that collaboration with or participation in GTOS is voluntary.

Lifetime of GTOS

GTOS should be conceived and designed as an "open-ended," long-term observing system.

GTOS will operate on the principle that long-term continuity of observations are essential for many environmental parameters and processes to establish quantitative rates of change.

Action items

Definition of GTOS

One of the critical action items for Working Group 1 (in collaboration with Working Groups 2 and 3) is to define carefully what GTOSshould be and what it should become. It was suggested that in this exercise, perhaps the initial task should be to set forth a strong rationale for why a global terrestrial observing system should, exist. At the same time, it may be appropriate to develop a clear statement of what it is not. Until a clear and full statement has been developed to explain why GTOS should exist and then to define what it should be and do, it will be difficult to establish a set of guidelines for the implementation and operation of GTOS.

Some of the ideas related to the definition of GTOS which were expressed at the Munich Workshop:

GTOS should have a coordinating function.
GTOS should serve as a catalyst in encouraging nations to share data and to use GTOS data.
GTOS should perform a function of harmonization of data from disparate sources.
GTOS should be an integrator of disparate kinds and sources of data.
GTOS should have a function of adding value to the data.
GTOS should be a distributor of credible, compatible, comparable, useful environmental data.
GTOS should be an evolving concept and program.
GTOS should have a unique role of its own.
GTOS should define its own "vision."

The basic concept is that GTOS should be an operational mechanism for catalyzing action, improving the availability of data, adding value to the data, and providing a forum for reaching agreements on harmonization. Among GTOS responsibilities should be the important roles of developing and promulgating a consistent information management framework, principles and protocols for systematic harmonized observation, and quality assurance methodologies to be followed by cooperating national and international centres in the GTOS network.

GTOS might also play an important role in nurturing collaboration and coordination of the various existing national and international monitoring systems and environmental study/monitoring sites. All such coordination activity should include collaborative definition of data/information requirements and data exchange procedures.

Objectives of GTOS

It is difficult to separate this action item from that of defining GTOS. However, it is essential to set forth a clear, limited set of objectives for GTOS before proceeding with other tasks of implementation. These objectives should serve as guidelines for determining the kinds and scope of other tasks to be undertaken by GTOS.

GTOS should be selective both in the issues it seeks to address and in the range of parameters or indicators it chooses to measure. Workshop participants cautioned against information overload and its ramifications for effective information management.

While the needs of the scientific community are important, it was affirmed that the primary aim of GTOS should be to accumulate and provide information necessary to make global decisions and give early warning of environmental concern to the international policy community.

Action Time Table

Having been announced and given publicity within the global change community, GTOS must design a time table and set specific goals for initiating and implementing the objectives and mission of GTOS.

Publicity

A concise, attractive colour brochure about GTOS, its objectives, and its approach to carrying out its mission should be prepared for educational purposes-informing appropriate international funding agencies, appropriate national and international political bodies, the global change community, the broader science community, and the public.

There is a need to "sell" the GTOS concept. An approach might be to indicate the value of GTOS information for meeting the requirements under post-Rio (Agenda 21) conventions. Selling the concept and the benefits of GTOS can help to overcome obstacles in data gathering from national sources, particularly in developing countries.

GTOS datasets

Even in the planning phase of GTOS, it is important to continue the refinement of the list of parameters and measurements (required, desirable, optional) of environmental-ecological data to be included in GTOS datasets. A shared conceptual model of processes which could lead to significant environmental changes at the global scale should guide the selection of a minimal set of variables to be observed. Where, how and when to measure these variables is another set of issues which must be addressed.

GTOS observing sites

Another GTOS planning task is to continue the inventory and examination of previously compiled lists and studies of ecological research sites (academic, research institutes, government research centers, other); determination of global sampling strategies which can best represent terrestrial ecosystem variability and global geographic distribution; selection and prioritization of a minimal list of observing sites for GTOS, taking into account technical, fiscal, and access feasibility and possibility. Not all variables should be measured according to a single sampling strategy. Global processes are not spatially uniform, nor is a single sampling methodology appropriate for all objectives. This question goes beyond simply site selection criteria (as addressed at Fontainebleu) to a fundamental consideration of sampling strategies and scaling considerations.

A definition of kinds of sites for terrestrial ecosystem observations must be considered concurrently with the consideration of sampling strategies (See Sampling Strategies under Research Priorities). One Working Group suggested three categories of GTOS observing sites:

– Intensive monitoring sites: a limited number of sites, monitoring a full range of parameters at the highest temporal resolution.

– Intermediate sites:

– Extensive sites: monitoring a few robust key parameters, requiring the minimum of resources and expertise; these key parameters must be those from which other required parameters for modelling can be inferred.

An important task to be implemented as early as possible is the development of collaborative agreements with terrestrial ecosystem research organizations to share appropriately documented environmental data for inclusion in GTOS datasets. An essential component of this task will be the development and definition of protocols and policies for the implementation of national and regional collaboration with GTOS.

A difficult but important task will be the identification of "gaps" – geographical and scientific – in the global terrestrial ecosystem arena. Once these gaps have been identified, a strategy for filling these gaps must be developed. Perhaps phase 1 in GTOS site selection should be a priority ranking of existing ecosystem study sites as GTOS candidate sites by means of (geo)statistical representativity analysis (of Chapters 2.2 and 3).

GTOS network

Attention is required early in the design and implementation of GTOS for the development of a directory for electronic access to and sharing of GTOS terrestrial ecosystem datasets.

GTOS should keep abreast of the evolving global communications and electronic networking systems, especially world wide web, and take advantage of these technologies as platforms for GTOS communications and as a tool for distributing GTOS data/information.

GTOS data centres

A GTOS coordinating centre, with a limited number of staff members, should be established. Special attention should be given to the provision of a centre where harmonization and integration of global scale data are performed. Successful operation of screening, analysis, and harmonization of environmental data presupposes skills in:

– data base management
– time series analysis

- spatial data analysis
- statistics/communications
- systems expertise
- appropriate disciplines
- coordination and administration

As GTOS evolves and its activities expand, a need for regional support centres may emerge. Such centres may be needed to provide a formal system for technical support, teaching and training. One of the important products of these centres will be value-added products generated from the analysis of data acquired from the GTOS network of study sites supplemented with global and site specific observations and measurements from other sources.

GTOS interface with other international observing systems

GTOS should develop a complete inventory of other global observing systems, establish contacts with leaders of these systems, and where possible and feasible, develop collaboration and data exchange with them.

Contacts and protocol for collaboration with GCOS (Global Climate Observing System) and GOOS (Global Ocean Observing System) have been established. Contacts and collaboration with other systems should be explored.

Demonstration of value of GTOS

It was recommended that a specific global issue or question be posed and addressed by GTOS to demonstrate that only GTOS information can provide the solution to the question. One suggestion was a question related to the amount of carbon, its rate of release from or intake into global soils, and the global variability of these processes. Other possible issues which might be addressed by GTOS were suggested.

It was emphasized by several participants in the Workshop that there is a rather urgent need to produce a credible, useful GTOS product as soon as possible and to demonstrate the utility of that product in rationalizing environmental policy decisions. One approach to demonstrating the value and utility of GTOS might be through pilot studies.

Funding of GTOS

Although the GTOS Ad Hoc Scienific and Technical Planning Group is co- sponsored by four agencies of the United Nations (FAO, UNEP, UNESCO, WMO) and the International Council of Scientific Unions, there is need even in the planning phase of GTOS to begin exploring funding possibilities for an operational GTOS.

Research priorities

Models for terrestrial ecosystems
There is a critical need to identify and inventory appropriate models and the data requirements for these models for which GTOS may become an important data source.

Research is needed on a comparative analysis of available models to identify minimum data sets required to run the models and to validate them.

Data requirements

A research task may be required to define and further refine the data requirements for the wide range of global and regional terrestrial modeling with a disparate set of objectives for quantifying global change and other phenomena within the Earth system. A related research issue is the determination of baselines of natural variability in space and time. This is a task fraught with great difficulty, but it is one which is extremely important for long-term monitoring of change. A research effort in paleo-ecological and historical reconstruction of ecosystems can contribute to the determination of baselines of natural variability.

An important consideration in GTOS' efforts to harmonize disparate data is research into the problem of controls and standards (reference materials) particularly relevant to the environmental sciences.

Quality assurance/quality control

A careful study must be made of established QA/QC methods used in ecosystem studies. The complexity of the range of parameters to be monitored over the many disparate terrestrial ecosystems of the Earth presents a formidable problem for developing credible QA/QC methods for GTOS. Appropriate research should be designed to address this issue.

Sampling strategies

A comparison of previously compiled lists of global sampling strategies should be made. There was agreement that GTOS will require several different sampling strategies. These strategies are not a menu from which to pick and choose. The only way to contain costs is to reduce the precision required, or to reduce the number of environmental aspects to be covered. Four recommended sampling strategies provide ideas which should be examined carefully and refined to serve the needs of GTOS.

Strategy One. Answering the question 'How much?' Variables for which absolute values with known precision are needed at the global scale require a global, unbiased, harmonised sampling strategy. Since most of these variables show great spatial variability, the sample number will generally have to be high, therefore, the cost per sample must be low. Sampling will be generally periodic and infrequent, since the rate of change is generally slow. The requirement for high n and no bias will generally preclude an approach based on an a priori set of sites. In general, this will need to be a centralised, top-down strategy.

Strategy Two. Answering the question 'Where?' Even a high-n sample at the global scale will not have sufficient resolution at the regional scale to apply targeted policy. For this a comprehensive and spatially continuous strategy, i.e. mapping, needs to be applied.

Strategy Three. Answering the question 'Why?' This requires detailed, intensive, continuous, site–based experimental and manipulative work. The number of sites need not be large, but

they will be individually expensive, and almost certainly nationally supported. The methods applied will often not be standardized. While the sites may be chosen to represent a particular ecosystem, they will not constitute a statistically valid or unbiased sample. Their purpose is to raise and test hypotheses, and build and validate
models.

Strategy Four. Giving early warning. It is highly unlikely that a random or representative sample of sites will be sensitive to change sufficiently ahead of time to be useful for early validation. Samples for this purpose should be placed in the path of postulated change: on steep and sensitive quadrents, in transition zones. By definition, these will not be representative. Their purpose is to confirm trends, validate hypotheses, and extrapolate predictions forward in time and space.

If GTOS integrates social and economic data into its system, further unique strategies will no doubt be required.

Extrapolation among different spatial and temporal scales

GTOS data sets will include point data as well as spatial and temporal data at a wide variety of scales. A variety of global sampling strategies are being considered. The credible integration of spatial and temporal data at many different scales is an important research problem. The difficulties of establishing quantitatively the relationships between different spatial and temporal scales should be examined for many components and processes in terrestrial ecosystems.

Metadata

Since a wide array of disparate data will be included as essential for describing the components, conditions and environmental processes at all GTOS sites, research must be conducted to develop a standard and minimal set of metadata and a compatible and comparable methodology for documenting these data.

Archiving

Since the concept for GTOS defines it as an open-ended long-range system, there will be a continuing challenge to the research community to determine what data are to be archived for potential use in the distant future. Related research must address the issues of the medium of storage and the assurance of capability to retrieve long-term archival data in the future.

Indicators of environmental change

It is extremely difficult to identify early indicators of environmental change. The lack of well defined and quantified benchline datasets adds to the difficulty. An important area of research is that of identifying and becoming knowledgable with previously developed screening methodologies and measurements of indicators of change and to expand on and refine them.

Quantifying rates of environmental change

Much research is yet to be done in the identification of biological indicators of environmental change. Since environmental change is a given and has been occurring through all time, GTOS should be concerned primarily with the identification or definition of unacceptable rates of change in the Earth system.

Integration of socioeconomic data with GTOS data

GTOS will operate in the context of concern for the human condition, the impacts of human presence and activities on the environment, and the impacts of environmental change on humans and their activities. GTOS is therefore challenged to encourage research on linking socioeconomic data to physical, chemical, and biological data related to environmental change.

Remote sensing systems

A broad array of aerospace sensors has evolved during the past three decades, and literally hundreds of thousands of space derived images of the Earth have been analyzed and/or archived. These remotely sensed images have been used in many different disciplines for characterizing components and conditions of the Earth surface (crop inventories, conditions of range lands, land degradation, biomass production, land use patterns and changes, surface cover, predictions of water yield from snow melt, drought, desertification, and other applications). With the great ranges in spectral, spatial and temporal scales now available through aerospace remote sensing technologies (past, present, future), remote sensing and relational spatial databases will play an increasingly important and essential role in routine monitoring of terrestrial ecosystems. These technologies have already made significant contributions in environmental research. However, there needs to be continuing support for research in the application of these technologies for monitoring, characterizing and assessing terrestrial ecosystems.

Capacity-building, technology transfer

From the beginning of GTOS activities, and especially during discussions with potential collaborating nations, attention should be given to the importance of the technological constraints to access to GTOS databases by developing nations. A researh priority should be given to exploration of the most rapid and efficient approach to the development of human resources and transfer of the technology in support of GTOS and all participating nations.

Distribution of data/information

Much research is needed to determine the most cost-effective and efficient methods to distribute GTOS products to policy-makers, other decision-makers, and the general community of GTOS data users.

Concluding statement

Among other objectives of the Munich Workshop on Harmonization of Environmental Data, it served as the first working meeting of the GTOS Working Group 1 on data management, access and harmonization.

Future deliberations and activities of WG1 should build upon and expand the conclusions and recommendations of the Munich Workshop under three specific areas-set of principles which should establish the nature, structure and operating guidelines of GTOS, a definition of GTOS and its objectives for rationalizing and justifying activities of an operational Global Terrestrial Observing System, and a set of recommendations for research priorities, specifically focused on GTOS data management, access and harmonization, in support of GTOS and related Earth observing systems.

3 Proposal for a Global Concept for Monitoring Terrestrial Ecosystems as a Basis for Harmonization of Environmental Monitoring

O. Fränzle, W. Haber and W. Schröder

Problem, Goal and Central Hypothesis

In Germany, there is scientific and political consensus about the need for environmental monitoring as an expression of the responsibility for protecting the environment. Nevertheless, there is still no final operational concept for developing such a system either in Germany or at the global scale. This article aims to describe the German experiences of environmental monitoring and to introduce these into the international discussion on harmonization of global environmental monitoring and the establishment of a Global Terrestrial Observing System – GTOS (HEAL et al. 1993; UNEP 1994).

As in Germany, in developing any practical environmental monitoring system at the global scale it is necessary to include existing measurement networks and to take account of the complex national and other areas of administration and jurisdiction. Both ecosystem theory (top down approach) and quality assurance and control (bottom up approach) are of critical importance for the harmonized use of existing national and international environmental monitoring systems.

The chances of realising the requirements when establishing a GTOS were investigated in close cooperation with the UNEP-HEM office and financial support from the Federal Ministry for Education, Science, Research and Technology (BMBF). Emphasis was on selection of appropriate areas of investigation in existing national and international networks (e.g. MAB Biosphere Reserves, IGBP, GCOS, GEMS/AIR, BAPMoN, EMEP; see TSAI-KOESTER 1994) from both the scientific (requirement profile) and practical (infrastructure, personnel etc.) points of view.

Environmental Monitoring in Germany

Definition and State of Development

Discussion of environmental monitoring requires – as with any scientific communication – as precise a definition of the concept as possible. Equally, the prerequisite for a *Scheme for the Development of Environmental Monitoring* is the harmonization of terminology in the sense of semantic precision and minimising redundancy. Established terminologies, however, should be taken into account as far as possible.

The Rat von Sachverständigen für Umweltfragen (Council of Environmental Advisors, SRU) published a first summary of the status and perspectives of environmental monitoring in Germany in 1990 (SRU 1990). This report is the starting point for the following which can be understood as an 'extrapolation' in the sense of the goal of this article.

The objects in the system should provide the starting point for developing a terminology for *environmental monitoring*. The SRU (1990, Secn. 7 to 9) takes this to be primarily the human environment. However, environmental monitoring 'cannot be restricted to the human environment alone' (ibid., Secn. 7). For this reason the term 'general environmental monitoring' should be used. This must 'cover the environment system as a whole, as well as the different environmental sectors or media … Thus we are referring to integrated environmental monitoring on the basis of a system or ecosystem' (ibid., Secn. 8). This leads to the term 'general ecological environmental monitoring'.

Parallel to this, the SRU also uses the following terms in its report (SRU 1990, Secn 9, p. 27; Secn. 69, p. 28): ecological environmental monitoring, integrating cross-sectoral environmental monitoring, ecosystem-based environmental monitoring, integrating long-term monitoring, and ecosystem monitoring. Since this takes place without any explicit semantic differentiation, it must be assumed that these terms are considered to be synonyms for 'general ecological environmental monitoring'. The subjects of environmental monitoring are 'regional ecological-economic systems'. These are composed of the three more narrowly differentiated systems 'nature', 'land use', and 'socio-economy' (ibid., Secn. 78).

Summarising the ideas of the SRU (1990) and systematising them for international discussion leads to the following: the subjects of environmental monitoring are ecological systems, i.e. ecosystems (HABER 1984, p. 54). These are composed of abiotic (air, water, soil) and biotic (humans, animals, plants) elements, which interact structurally and functionally.This concept of an ecosystem is not restricted a priori to a particular spatial scale (TANSLEY 1935, cited in HABER 1993, p. 15). Thus the planet earth can be viewed as an ecosystem, as can be a forest or a hedgerow. The spatial and temporal resolution used for ecosystem monitoring should be chosen according to the spatial and temporal differentiation of the elements and structure and function of the system. Furthermore, the spatial and temporal detail of the monitoring should be appropriate for the intended application of the results. Finally, in addition to these more technical requirements, there will be some need to take into account the financial and personnel resources.

The aim of ecosystem monitoring should be to understand the compartments, structures and functions in interaction. Thus it would be inappropriate to exclude man from environmental monitoring. Man is influenced both as an individual and as a species by his environment. Equally man affects his environment. Thus ecosystems can be characterised not only on the basis of structural and functional characteristics, but also in terms of anthropogenic influence, e.g. through hemerobic status, use or culture gradients or stability units (FRÄNZLE 1993; GRIMM 1994; GRIMM et al. 1992; HABER 1993, pp. 71–75; JALAS 1955).

The above means that for the terminology the term 'general ecological environmental monitoring' and the synonyms named above can be replaced by the term 'ecosystem monitoring' without any loss of information. Furthermore, apart from the question of terminology, it is necessary that environmental monitoring

– is the system in which environmental research in the natural and social sciences and humanities should be embedded,

– should be designed taking into account the possible applications of results in environmental and nature protection as well as environmental law and politics,

– is supported by theoretical scientific and statistically well-grounded theories of measurement, and

- leads to data based on the above which are quantified as far as possible through quality assurance and control.

Against this background, we should discuss whether the term environmental monitoring (ecosystem monitoring) should be expanded by 'assessment'. This makes sense when, following ELLENBERG et al. (1978),

- we consider ecosystem monitoring as the basis for an environmental information and evaluation system and

- define it with regard to environmental politics and planning and nature protection, i.e. in terms of actions to be taken.

This is because rational action is based on decisions which are themselves preceded by an evaluation of possible alternatives. An assessment is a communication about preferential relationships between real objects or facts which prepares for action. For this a factual model (a virtual image of a past, present or future piece of reality) is linked to one or more value systems. The elements of this model are defined in terms of relative importance by means of decisions (value judgements). Thus in ecosystem assessment empirical-descriptive scientific propositions are related to normative propositions in such a way that resultant actions in the fields of environmental and nature protection and environmental politics are linked to certain measurement values.

This type of comprehensive application-oriented environmental monitoring and assessment does not yet exist in Germany. Up to now, both the methodology and contents of the basic concept given by ELLENBERG et al. (1978) have been developed in a more concrete fashion in predominantly scientific projects supported by both the Federal Ministry for the Environment, Nature Protection and Nuclear Safety (BMU), the Federal Ministry for Education, Science, Research and Technology (BMBF) and the Länder authorities. These include amongst others:

- the joint projects on ecosystem research in Bayreuth, Berchtesgaden, Göttingen, Kiel, Leipzig-Halle and Munich;

- the development of a concept, and testing of a planning-oriented, regionalised environmental monitoring (FRÄNZLE et al. 1992); environmental monitoring in natural forest reserves (PROJEKTGRUPPE NATURWALDRESERVATE 1993), the immission effects register of Baden-Württemberg (ZIMMERMANN 1991);

- the development, testing and setting up of an *environmental specimen bank* as an instrument for the assessment of the actual state and future trends of environmental loads, and as retrospective view of previous conditions of ecosystems.

Other environmental monitoring projects in Germany are mainly sectorally oriented, i.e. concentrating on specific environmental media. The SRU (1990, Secn. 11) sees a major shortcoming in these activities to be the general lack of any harmonisation of methodology, content or organisation. In addition, sectoral monitoring which is not linked to other measurement networks impedes the interpretation of data in terms of ecosystem function and structure. A system of ecosystem monitoring and assessment which is to be developed in accordance with the above requirements should be so structured that it both makes use of, and supplements, existing sectoral monitoring activities (FRÄNZLE et al.1992, p.7; SCHÖNTHALER et al. 1994, p. 38).

Relationships to Environmental Research and Tasks

Both in a national and an international framework, the tasks of environmental monitoring should be defined as a contribution to environmental research. Environmental research means those activities aimed at the intersubjective comprehensible production of objective, reliable, and valid basic knowledge and extensive information about ecosystems (which in the following is taken to mean terrestrial ecosystems).

Nowadays environmental research in Germany is undertaken both within the natural and social sciences and the humanities disciplines. As discussed above, efforts should be made to integrate these different approaches in the field of environmental monitoring and assessment (SCHRÖDER & DASCHKEIT 1994; KEUNE et al. 1993) (Fig. 3.1). The main elements of the environmental monitoring and assessment system as proposed by ELLENBERG et al. (1978) have now been tested in Germany. Three subsystems fulfil essentially the following functions:

– *Ecosystem* research is basic research in the biological and geoscientific disciplines, aiming to increase understanding of the structure and function of ecosystems and their carrying capacity. Closely linked to this, both spatially, methodologically and organisationally, is

– the *environmental specimen bank*. It serves the storage of environmental specimens from the study areas of ecosystem research and monitoring. The aim is both documentation and spatial and comparative retrospective analysis.

– *Environmental monitoring* (in the limited sense) serves the recording within an area of the status and development of representative ecosystem types and their relevant compartments, with the aim of contributing to environmental assessment and planning. Environmental assessment necessitates both comparison of the state of the environment in time and space as well as the relation to environmental quality goals and environmental standards developed in accordance with these (HABER 1989).

The tasks of environmental monitoring (ecosystem monitoring) within the framework of the three-part concept of environmental monitoring and assessment (see above) can be summarised as follows:

– Environmental monitoring fulfils a *confirmative function for hypotheses,* for example when it makes available data on the general spatio-temporal applicability of experimentally verified hypotheses on the causes of forest damage by pollutants (SCHRÖDER 1994b).

– Environmental monitoring can have an *explorative function for hypotheses* when it makes available data on the spatio-temporal coincidence of certain environmental factors and ecosystem states which lead to the instigation of experimental investigations in ecosystem research about corresponding cause-effect relationships.

In addition to these deductive and inductive functions which are more a part of environmental research, environmental monitoring can also fulfil
– planning tasks and
– control functions in environmental law.

Thus specific areas can be selected as protected reserves on the basis of long-term ecosystem monitoring, and the effect of human impact on ecosystems can be evaluated. In addition to

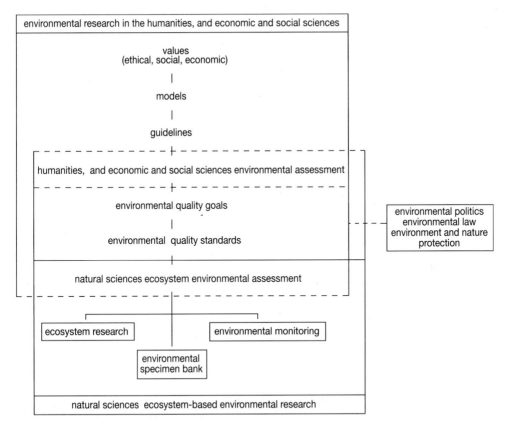

Fig. 3.1:
Position of Envionmental Monitoring within the System of Environmental Research (Draft: W. Schröder)

these precautionary tasks, data from environmental monitoring networks also serve to provide evidence and analysis of damage, and thus contribute to conviction under environmental law.

Harmonisation of Environmental Monitoring and Assessment in National and International Cooperation

Requirements

As described above, data from environmental monitoring provide an important basis for environmental politics, environmental law and environmental protection activities (HABER 1993). This is true at all scales of environmental monitoring, from local to global. This hierarchical concept (SRU 1990, p. 30) often requires the generalisation of large-scale data (1:100 to 1:10 000), i.e. data collected at a point or within a small area, to larger areas. Such extrapolation and aggregation is only meaningful if the original data at the most detailed level are gathered in a harmonized fashion. For this it is necessary to have agreement about the main processes of gathering data. These include necessarily the

- aim of the project,
- scientific terminology/nomenclature,
- selection of the objects to be considered (here: ecosystems),
- selection of characteristics to be observed on the objects (here: functional and structural indicators),
- method of monitoring (from the purely visual to monitoring supported by instruments, see below definition of 'measure'), and
- evaluation methods (statistical procedures, simulation models, ...).

Harmonisation does not imply that all the above processes must be identical, i.e. standardized. The possibility of realising this 'ideal' is inversely proportional to the number of monitoring systems (measuring stations), and at a global scale the problem increases directly with the degree of difference in the development of infrastructures. Harmonisation is much more concerned with achieving comparability of monitoring by compilation or development of interfaces within (horizontal) and between (vertical) different levels of monitoring. This does not exclude the possibility of standardization - where possible or necessary. Thus the harmonisation of global environmental monitoring should concentrate on the above and those given below in more concrete examples.

Obtaining information can proceed in different ways, for example through collection of data. Data gathering is based on measurement activities (i.e. monitoring) which are determined both by a system of measurement and a protocol. Measurement is the assignment of numbers to characteristics of an object whilst maintaining an analogous relationship which is based on the system and protocol. The meaningfulness of such a representation depends on (SCHRÖDER 1994a, pp. 6, 7; UNEP & WHO 1994a, b):

(a) the representativeness of the elements (objects) in the investigated sample for the whole (truth of representation 'sample – whole'),

(b) the choice for testing the hypothesis of relevant object characteristics and methods with whose help particular characteristics of the sample elements will be measured (possibility of making a statement and comparability of investigation methods),

(c) the main quality criteria for measurement procedures: objectivity, reliability and validity (truth of representation 'investigated sample characteristic – measurement value' and comparability of the investigation results).

If these criteria are not fulfilled at the lowest level of environmental monitoring then they will not be fulfilled at any other level. Thus harmonization efforts must focus on these criteria. Quality assurance systems are a prerequisite for their realisation and observance. Quality assurance systems must be integrated as a permanent part of programmes for environmental monitoring. It must start with the selection of the object to be investigated and continue through to the documentation of results (Fig. 3.2). Thus it covers *preparatory quality control (selection of representative objects for investigation and decisions on measurement* methods), quality control of measurements (internal laboratory routine quality control and external quality control), and *quality control of data* (testing for plausibility and spatio-temporal meaningfulness of the measurement results).

If the results of investigations are to be representative, i.e. meaningful beyond the spatial and temporal limits of the sample from which they are taken, then the following requirements are relevant (SCHRÖDER 1994a, p. 5):

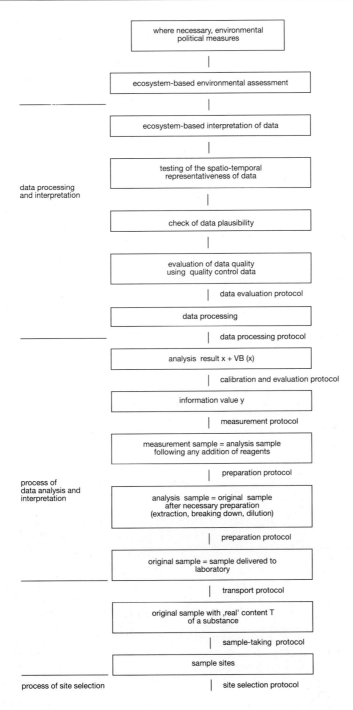

Fig. 3.2:
Gathering Data in Environmental Monitoring Using the Example of Chemical Analytical Measurement
(Draft: W. Schröder)

1. The sample must be a reduced image of the whole (i.e. the global ecosystem inventory), both in terms of heterogeneity and in relation to the functional and structural representativeness of the relevant variables for testing hypotheses (see above a,b).
2. The elements of the sample must be defined.
3. It must be possible to define the whole empirically.
4. The method used to select the sample elements must be describable and fulfil criterion 1.

If it is not possible to fulfil the criteria (1) to (4) then it is absolutely necessary for the evaluation of the results to describe and justify the criteria and processes used to select the samples and methods.

In contrast with standardized analytical measurement methods (DIN, ISO, etc.) only few generally recognised standards exist to control for possible errors in measurement processes, e.g. in laboratory analyses. The aims of measurement quality assurance are:

– statements on the achieved precision and accuracy of the analytical results and
– ensuring the reliability of analytical results, i.e. safeguarding the accuracy and precision over the period of monitoring.

Thus measurement results are comparable when they are properly and reliably obtained and controlled by methods with the same potential significance of meaning. The spatio-temporal representativeness of data which are quality-controlled in this fashion should be characterised as far as possible by statistical methods.

The quality control data must be comprehensively documented and made available to (central) environmental data bases. According to the above, it is necessary in the case of an analytical chemical measurement, for example, to provide details of: criteria for the selection of the sample and the methods for sample preparation and analysis; sample-taking methods; number of objects investigated; analytical apparatus; working range and limits of detection of the analytical system; precision and accuracy of measurements; level of significance and likelihood of error in measurement comparisons, correlation and regression calculations etc.

For the comparability of environmental research results and the assessment of spatial and temporal trends, it is necessary to have minimum programmes of investigation. These should be planned or expanded and justified in relation to specific types of ecosystems or questions. In order to evaluate data from environmental monitoring it is necessary, in addition to fulfilling the above requirements, that the concepts and decisions made are transparent and comprehensible.

Recommendations

The monitoring of terrestrial ecosystems on a global scale is indispensable with regard to the focus of international environmental politics on sustainable development. German environmental politics and German environmental research also agree with this aim and support its realisation. A part of this is the participation in the preparation and implementation of a *Global Terrestrial Observing System* (GTOS).

The following approach should ensure that the national contribution to GTOS is as effective and useful as possible:

1. A workplan with a defined time-scale for the development and planning of GTOS should be drawn up.

2. For this, the aims and functions of GTOS should be clearly defined. This is of particular importance for the question of whether GTOS should play more of a catalytic function between existing environmental monitoring programmes, or should be actively involved in the collection of data on the structure and function of ecosystems.

3. In order to clarify the relationship between 'catalytic' and 'active' roles, the geographic and scientific gaps within the existing environmental monitoring networks should be identified. This task is of highest priority. It should be carried out within the framework of a feasibility study. Relevant experience (AUSTIN et al. 1994; FRÄNZLE et al. 1992; DASCHKEIT et al. 1993; KOTHE & SCHMIDT 1994; SCHMIDT et al. 1995) has shown that for this it is necessary to combine data on the following, amongst others, within a geographic information system:
   a) potentially suitable research and monitoring areas (geographic position, altitude, biogeographic province, flora, fauna, climate, hydrology, geology, ...),
   b) research topics and research plans at these sites,
   c) measurement values (including quality control data),
   d) apparatus available,
   e) data and sample archivating methods and capacity,
   f) scientific disciplines participating,
   g) time of investigations (from ... to).

   This information should then be evaluated in terms of:
   – scientific basis (methods, interdisciplinarity, measurements, ...),
   – geographic completeness,
   – representativeness of investigated ecosystems (types),
   – possibility and realisation of contribution of the relevant site to the ecological monitoring and assessment of terrestrial ecosystems,
   – ...
   The evaluation should result in:
   – a priority listing for possible sites for GTOS and
   – proposals for further sites where necessary.

4. The aims, functions and measures for implementation of GTOS should be laid down in accordance with these results.

At the international level some of the above tasks are particularly urgent. These include, for example, the definition of a basic programme of measurements on a continental and global scale. This presupposes, however, that the thematic, methodological and infrastructural homogeneity/heterogeneity of *environmental monitoring* has been agreed and evaluated on a global scale in the sense of the recommended approach. The authors feel that the Biosphere Reserves of the UNESCO programme 'Man and the Biosphere' (MAB) would be particularly appropriate for this on the basis of the integrative, i.e. natural, cultural and social scientific research and management approaches developed (ERDMANN & NAUBER 1990; NATURE & RESOURCES 1993).

Even though the MAB Biosphere Reserves have not up to now been established primarily with a view to environmental monitoring they can in general be used for this purpose; this aim should be recognised in the selection of further Biosphere Reserves. Particularly these areas which in contrast to nearly all other types of reserves are treated internationally according to the same criteria, are predestined as a location for the global harmonisation of environmental monitoring within the framework of long-term development cooperation.

References

AUSTIN, M. P.; MARGULES, Ch. R. (1994): Die Bewertung der Repräsentanz. In: USHER, M. B.; ERZ, W. (Hrsg.): Erfassen und Bewerten im Naturschutz. Probleme, Methoden, Beispiele. – Heidelberg, Wiesbaden, S. 48–65

DASCHKEIT, A.; KOTHE, P.; SCHRÖDER, W. (1993): Repräsentanzanalyse zur Auswahl von Bodendauerbeobachtungsflächen in Brandenburg. – Kiel (im Auftrag des Landesumweltamtes Brandenburg)

ELLENBERG, H.; FRÄNZLE, O.; MÜLLER, P. (1978): Ökosystemforschung im Hinblick auf Umweltpolitik und Entwicklungsplanung. – Berlin, Kiel (Umweltforschungsplan des Bundesministers des Innern. Forschungsbericht 78-101 04 005, im Auftrag des Umweltbundesamtes)

ERDMANN, K.-H.; NAUBER, J. (1990): Biosphären-Reservate – Ein zentrales Element des UNESCO-Programms "Der Mensch und die Biosphäre" (MAB). In: Natur und Landschaft, 65 (10), S. 479–483

FRÄNZLE, O. (1993): Contaminants in terrestrial ecosystems. – Berlin (...)

FRÄNZLE, O.; RUDOLPH, H.; DÖRRE, U. (1992): Erarbeitung und Erprobung einer Konzeption fur die ökologisch orientierte Planung auf der Grundlage der regionalisierenden Umweltbeobachtung am Beispiel Schleswig-Holsteins. – Berlin (UBA-Texte 20/92)

GRIMM, V. (1994): Stabilitätskonzepte in der Ökologie: Terminologie, Anwendbarkeit und Bedeutung für die ökologische Modellierung. – Diss., Marburg

GRIMM, V.; SCHMIDT, E.; WISSEL, C. (1992): On the application of stability concepts in ecology. In: Ecological Modelling, 63, pp. 143–161

HABER, W. (1984): Erwartungen und Ansprüche an die ökologische Forschung. In: MAB-Mitteilungen, Bd. 21, S. 54–72

HABER, W. (1989): Notwendigkeit und Funktion von Umweltstandards. In: Jahrbuch 1988 der Akademie der Wissenschaften zu Berlin, S. 263–295

HABER, W. (1993): Ökologische Grundlagen des Umweltschutzes. – Bonn (Umweltschutz – Grundlagen und Praxis; Bd. 1)

HEAL, W. O.; MENAUT, J.-C.; STEFFEN, W. L. (eds.) (1993): Towards a Global Terrestrial Observing System: Detecting and monitoring change in terrestrial ecosystems. – Paris, Stockholm (MAB Digest 14 and IGBP Global Change Report 26)

JALAS, J. (1955): Hemerobe und hemerochore Pflanzenarten. Ein terminologischer Reformversuch. In: Acta Societas pro Fauna et Flora Fennica, 72 (1), S. 1–15

KEUNE, H.; SCHRÖDER, W.; VETTER, L. (1993): Globale Integrierte Umweltbeobachtung und -bewertung. In: MAB-Mitteilungen, 37, S. 51–59

KOTHE, P.; SCHMIDT, R. (1994): Nachbarschaftsanalytische Ausweisung repräsentativer Bodendauerbeobachtungsflächen in Brandenburg. In: SCHRÖDER, W.; VETTER, L.; FRANZLE, O. (Hrsg.): Neuere statistische Verfahren und Modellbildung in der Geoökologie. – Braunschweig, Wiesbaden, S. 226–237

NATURE & RESOURCES (1993): Biosphere reserves - theory and practice. Special issue, Vol. 29, No. 1–4

PROJEKTGRUPPE NATURWALDRESERVATE (1993): Empfehlungen fur die Einrichtung und Betreuung von Naturwaldreservaten in Deutschland. In: Forstarchiv, 64, S. 122–129

SCHMIDT, R.; SCHRÖDER, W.; TAPKENHINRICHS, M. (1995): Soil data for models and resource management at different scales. In: Transactions 15th World Congress of Soil Science, Acapulco (Mexico), Vol. 6a, pp. 58–66

SCHÖNTHALER, K.; KERNER, H.-F.; KÖPPEL, J.; SPANDAU, L. (1994): Konzeption für eine ökosystemare Umweltbeobachtung – Pilotprojekt fur Biosphärenreservate. – Berlin (Umweltforschungsplan des Bundesministers für Umwelt, Natur und Reaktorsicherheit. Forschungsbericht 101 04 0404/08 in der Fassung vom 30.04.1994, im Auftrag des Umweltbundesamtes)

SCHRÖDER, W. (1994a): Erkenntnisgewinnung, Hypothesenbildung und Statistik. In: SCHRÖDER, W.; VETTER, L.; FRÄNZLE, O. (Hrsg.): Neuere statistische Verfahren und Modellbildung in der Geoökologie. – Braunschweig, Wiesbaden, S. 1–15

SCHRÖDER, W. (1994b): CHAID-Analyse des Bedingungsgefüges von Waldschäden. In: SCHRÖDER, W.; VETTER, L.; FRÄNZLE, O. (Hrsg.): Neuere statistische Verfahren und Modellbildung in der Geoökologie. – Braunschweig, Wiesbaden, S. 195–223

SCHRÖDER, W.; DASCHKEIT, A. (1994): Ökosystemare Umweltbewertung. Thesen zum Anforderungsprofil zukünftiger Umweltforschung. In: Umweltwissenschaften und Schadstoff-Forschung – Zeitschrift für Umweltchemie und Ökotoxikologie, 6 (3), S. 139–144

SRU (Der Rat von Sachverständigen für Umweltfragen) (1990): Allgemeine Ökologische Umweltbeobachtung. – Stuttgart

TANSLEY, A.G. (1935): The use and abuse of vegetational concepts and terms. In: Ecology, 16, pp. 284–307

TSAI-KOESTER, L.-H. (1994): A survey of environmental and information management programmes of international organizations. – 3rd ed., Munich

UNEP (United Nations Environment Programme) (1994): Report of the first meeting of the ad hoc scientific and technical planning group for a Global Terrestrial Observing System (GTOS) 13–17 December, 1993, Geneva, Switzerland. – Nairobi (UNEP/GEMS Report Series No. 24)

UNEP & WMO (United Nations Environment Programme and World Meteorological Organization) (1994): Quality assurance in urban air quality monitoring. – Nairobi (GEMS/AIR Methodology Review Handbook Series, Vol. 1: WMO/EOS/94.1, UNEP/GEMS/94.A.2)

UNEP & WMO (United Nations Environment Programme and World Meteorological Organization) (1994): Primary standard calibration methods and network intercalibrations for air quality monitoring.- Nairobi (GEMS/AIR Methodology Review Handbook Series, Vol. 1: WMO/EOS/94.2, UNEP/GEMS/94.A.3)

ZIMMERMANN, R.-D. (1991): Ökologisches Wirkungskataster. In: VDI-Berichte, 901, S. 61–71

4 Summary

H. Keune, P. Mandry, W. Schröder and O. Fränzle

In order to fulfil our responsibility for the global environment and for sustainable development throughout the world, it will be necessary to develop a global understanding of the environment and of the consequences of actions based on such understanding. To think globally and to act locally is the challenge we all have to face.

The need to provide the factual basis for such a global understanding, and for the required local and regional action led the climatological and oceanographic research communities to join efforts and plan the establishment of the Global Climate Observing System (GCOS) and the Global Ocean Observing System (GOOS). Subsequent recognition of the lack of a concerted effort to monitor the terrestrial contribution to global change and its effects, led five major international organizations, UNEP, UNESCO, WMO, FAO and ICSU to cooperate in planning and preparing a scientifically based design for a proposed Global Terrestrial Observing System (GTOS). The GTOS activities are intended to cover all aspects of terrestrial environmental monitoring and assessment. The major new aspect in the design of these observing systems is the necessity for a cross-sectoral approach, that is the integration of information from a wide variety of disciplines, and the need to use data from a wide variety of different sources and programmes. These data must be comparable and compatible of different sources and programmes if their combination on integrated assessments is to be justified and the information obtained meaningful. Solving this problem of data (and methodology) harmonization will thus be central to the success of the global observing systems. These systems have both a pressing need for harmonization and provide a possible means for developing harmonization on a global scale.

As they emerge, the global environmental observing systems will undoubtedly depend on the strengths of relevant international, regional, national and local environmental monitoring programmes. Clearly, not all of the available environmental data will contribute usefully to a global observing system, but since the information for global interpretation is generated by integration and aggregation of single observations, all data may have some relevance. Accessible data, however, have first to be checked for their comparability and compatibility with other existing data. The integration and aggregation which is then necessary must be carried out in such a way that the comparability and compatibility is further enhanced. Harmonization is required at every step in this process.

The UNEP Harmonization of Environmental Measurement (HEM) office has been considering problems of harmonization in environmental monitoring, observation and information management for more than four years. HEM has identified two prerequisites for genuinely useful cross-sectoral harmonization activities: a conceptual framework for monitoring activities, and a comprehensive and consistent catalogue of information on the contributors to the overall endevour.

HEM considers that GTOS could, or should, provide a focus for global cross-sectoral harmonization within the area of terrestrial environmental monitoring, as well as providing a real practical basis for developing a framework for further harmonization efforts. Equally, one of the keys to the success of GTOS will be the development of effective integrated measures for harmonization of data collection, integration and assessment activities for the network itself. All of this is in the centre of the activities of the HEM office.

A useful starting point for considering the development of a conceptual framework for environmental monitoring and research for GTOS can be found by examining the examples presented in a number of programmes at a local scale in the fields of integrated ecosystem monitoring and research. Based on the experience gained from these approaches, and in order to support the further definition and development of GTOS, HEM wants to identify and elaborate the critical harmonization and standardization requirements with respect to environmental monitoring and assessment. This was the task of the workshop described in this book. The workshop, held on 7–10 June 1994 in Munich, Germany, brought together experts from a broad cross-section of disciplines involved in environmental monitoring programmes. The workshop focused on issues related to harmonization of environmental data, especially at the global and regional scales, in support of the proposed Global Terrestrial Observing System (GTOS). The workshop was based on the experience of previous studies of monitoring and assessment of terrestrial ecosystems. Discussions centred on the recognition of and removal of a wide range of constraints to management, access and harmonization of environmental data for the proposed GTOS.

From the chapters in this volume, it is clear that a GTOS is urgently needed if we are to "fully" understand the terrestrial contribution to global change. The success of a GTOS, however, will depend on obtaining reliable and compatible information from the many hundreds of existing environmental monitoring and research programmes and networks. The urgent need for reliable and compatible information in turn increases the importance of harmonization of techniques and practices of measurement world-wide. This scenario shows the complexity of the problems involved in defining GTOS and in the harmonization of environmental data.

The chapters of this book address the specific issues of the constraints which exist to the management, access and harmonization of environmental data, the need for and benefits of GTOS, and the practical implications of harmonization for a global network. They include many examples of environmental monitoring at global, regional and national levels and together they bring a spectrum of experiences and points of view that will be helpful for the further development of GTOS. A few contributions provide insight into specific thematic issues from the wide range of areas which will converge in GTOS. The latter contributions relate to topics such as extraction of environmental information from complex datasets; model-supported synthesis, evaluation and application of environmental information; integrated assessment; estimation of gas fluxes between atmosphere and terrestrial ecosystems; environmental economic accounting to support decision-making; and standard reference data and policy assistance for human health. A brief summary is given in the following.

The rationale, aim and objectives of the book are described in Chapter 1. In Chapter 2.1.1, E. F. Roots discusses the historical background of environmental information requirements, data collection and data management. He suggests that the increasing amount of existing environmental information and increasing number of sophisticated information-rich monitoring systems could either contribute to an "information highway" of the future or could lead us into a maze. Specifically for GTOS, the direction taken will depend to a great extent on how successfully scientists "harmonize" the information to make it coherent and meet the respective needs of the diverse kinds of decision-makers in different components of the society.

As outlined by A. L. Dahl in Chapter 2.1.2, the UN system-wide Earthwatch is a mechanism to stimulate the UN system, in collaboration with governments and the international scientific community, to gather, integrate and analyze data and information as the basis for comprehensive assessments of environmental issues, early warning of environment threats and the identification by the international community of policy options and management responses for achieving sustainability. Earthwatch has identified a particular weakness in terrestrial assessments and supports the establishment of a global terrestrial observing system.

In Chapter 2.1.3, M. F. Baumgardner describes the developments which led to the planning of a global terrestrial observing system, and the current state of the proposed GTOS. He emphasises the importance of interaction of GTOS with the climate and ocean observing systems to ensure the production of harmonized data across the whole spectrum of observations. Without this it will not be possible to develop a consistent view of the environment world-wide.

Successful development of a global terrestrial observing system will require a clear definition of GTOS product users, identification of the resources and issues of concern, establishment of goals and objectives, and the formulation of a conceptual approach. In Chapter 2.1.4, H. Narjisse identifies national information needs and the requirements of developing countries for specific issues such as soil erosion, deforestation, loss of biodiversity, pollution in urban areas, drought and desertification; as well as the main constraints facing the development of environmental activities in these countries. He describes mechanisms to stimulate the participation of deve-loping countries in GTOS in order that they can both benefit from GTOS products and contribute to the development of the system. Among these incentives, he refers to assistance in policy formulation, institution building, human resource development and technology transfer. He emphasises the need to secure long-term commitment of developing countries to monitoring activities.

In Chapter 2.1.5, H. Keune and P. Mandry outline the harmonization requirements involved throughout the process from environmental observations to policy setting, i.e. from data collection to development of information by data integration and aggregation, to formation of environmental knowledge as the common basis for decision-making. They suggest that GTOS could provide a focus for global cross-sectoral harmonization within the area of environmental monitoring, as well as providing a real practical basis for developing a framework for further harmonization efforts.

Observing produces data which have to be of defined quality to know the extent to which they represent they "true" reflections of the observed reality. Today there is a great amount of environmental data available. But often these data are of unknown quality, and many relevant data are missing. This is especially true for terrestrial ecosystems on the global scale. To improve the situation, in Chapter 2.2 W. Schröder and O. Fränzle outline a conceptual framework for environmental monitoring and assessment as a basis for harmonization needs, using the example of the Ecosystem Research in the Bornhöved Lake District, Germany. It should be based on ecosystem research and consist of procedures for: (i) assessment of environmental data, (ii) data collection by means of environmental monitoring and research, (iii) integration and aggregation of environmental data by means of statistical methods, simulation models, and/or scenario techniques.

Data management and analysis in ocean sciences within national and international scientific programmes is discussed by D. Kohnke in Chapter 2.3.1. Specific international mechanisms and regulations exist to monitor the flow of data from data collection to data exchange. The

Global Ocean Observing System, GOOS, is being planned as an internationally coordinated system for systematic operational data collection, data analysis, exchange of data information, and technology development and transfer. GOOS will use a globally-coordinated, scientifically-based strategy to allow for monitoring and subsequent prediction of environmental changes globally, regionally and nationally.

In Chapter 2.3.2, V. Mohnen discusses quality assurance/quality control (QA/QC) as a central issue for harmonization at the level of data collection and management, and the approach used in the Global Atmosphere Watch (GAW) programme of the World Meteorological Organization. The overall role of GAW is to supply data and information of known quality which is necessary for assessment of status and trends in the global atmospheric environment. Quality Assurance/Science Activity Centres are being established to provide the operational mechanism for harmonization and coordination of GAW data gathering. Specific duties of these centres include preparation of quality assurance plans, quality assurance support, management review of data and quality control products, training, and information dissemination.

The Arctic Monitoring and Assessment Programme (AMAP) is a regional programme concerned with environmental monitoring of the Arctic environment. The main objective of AMAP is to monitor the levels of pollutants and assess the effects of anthropogenic pollution in the Arctic. Monitoring activities related to specific parameters in different media and to biological effects, and the interactions with climate change and human health, are outlined by L. O. Reiersen in Chapter 2.3.3. The need for internationally recommended quality assurance methods and procedures has been recognised for a series of components.

Integrated environmental monitoring refers to the combined assessment of a wide variety of measurements from different disciplines but related to a single environmental space in order to determine cause-effect or other relationships, and obtain comprehensive assessments of status and trends. Integrated monitoring is particularly important in identifying interface relationships, e.g. land-atmosphere or landwater, and in assessing feedback responses. The Environmental Change Network (ECN) of the United Kingdom is an integrated monitoring network of representative sites which links physical, biological and chemical variables. In Chapter 2.3.4, A.M. Lane describes the ECN approach with respect to criteria selection of sites and core measurements, collection and management of data, and quality assurance procedures.

In Chapter 2.3.5, K. Peter and P. Knoepfel describe a Swiss approach to integrated eco-system-related environmental observation based on the ecosonde concept. The measurement design developed following the implementation of the observation program in a specific observation area, is known as the ecosonde. The ecosonde is a highly sensitive, areaspecific measurement instrument to be used in integrated environmental observation.

Diverse topics relating to the development of GTOS are described by different authors. The requirements of GTOS relating to harmonization and operational and financial issues are outlined by J. Nauber (Chapter 2.3.6). L. A. Ogallo (Chapter 2.3.7) discusses specific user requirements for GTOS concentrating on the needs of the developing countries. In Chapter 2.3.8, Y. A. Pykh describes the experience of the Russian Centre for International Environmental Cooperation (INENCO) with modelling of terrestrial ecosystem dynamics, and emphasises the need to establish a Russian environmental database to compile existing terrestrial information and integrated into other international systems. GTOS will build on existing environmental programmes, and in Chapter 2.3.9, J. Cohen considers the necessary steps for establishing national/regional data centres to be coordinated by GTOS.

In Chapter 2.4, I. K. Crain identifies some of the constraints to the extraction of information from environmental datasets, i.e. the use of complex databases, difficulty to integrate data from different database technologies, undetectable linkages because of non-coded information, lack of reliable Geographical Information System (GIS) models, and the lack of effective methods of presentation of information to decision-makers with integrated "views". He suggests ways of breaking these constraints, namely to encourage the development of spatial support software which incorporates "decision-maker views", to develop, document and utilise meaningful natural resource models which also relate human factors to environmental change, and to educate decision makers on the availability and use of existing tools.

H. F. Kerner proposes a multidimensional, multilevel and multifunctional interdisciplinary concept as the basis for integrated ecosystem monitoring (Chapter 2.5). The concept is a hierarchical model. It focuses on the level of ecotopes which represent different gradients of ecosystems which are aggregated and integrated into a generalised ecosystem model. The model is based on the internal balance of matter and energy within the ecotopes and their external relations in space. Spatial evaluation using GIS allows an integrated analysis and impact assessment of the environment at the landscape level including human activities.

An integrated assessment concept for the storage and retrieval of data and models relevant for impact assessment studies at the ecosystem and landscape levels is described by R. J. M. Lenz and co-authors in Chapter 2.6 The concept is based on databases including GIS, and models, as well as integration methods like (geo)statistics, and environmental classification. A database is being designed to offer information on ecological models and their data requirements so that users can evaluate the quality of the models. A case study in integrated assessment for landscape diversity and land-use planning is discussed.

In Chapter 2.7, U. Dämmgen and co-authors describe a model for estimating flux densities of air-borne matter, and recommend measures to meet the needs for the application of these models for monitoring purposes.

The need to understand the links between status and trends in the environment and economic statistics has been widely recognised. As discussed by W. Radermacher in Chapter 2.8, environmental economic accounting aims to show in statistical terms which natural resources are used, consumed, depleted, or destroyed by activities (production/consumption) during a defined period, and what expenditure is necessary for countermeasures. These calculations are based on the process of creating 'value-added' as reflected in economic statistics. A concept showing the working area of environmental economic accounting is described.

In order to provide standard reference data and policy assistance systems for global health evolution, C. Greiner and co-authors (Chapter 2.9) propose an "optimistic" modelling framework which separates scenarios of politically important future input variables for models from correspondingly optimistic predictions by using simple models with a monotonic structure.

The main results of the workshop are summarised by E. F. Roots and M. F. Baumgardner in Chapter 2.10. E.F. Roots stresses the need for visions and goals of GTOS, and refers to the clients that GTOS aims at; he also indicates the major constraints involved in the development and implementation of a coordinated environmental observing system which includes socio-economic variables and is global in scope and science-driven but politically supported. The conclusions of the meeting based on the reports of the workshop working groups are summarised by M.F. Baumgardner. He outlines a series of guiding principles, action items and research priorities as a basis for the further planning and development of GTOS.

In Chapter 3, O. Fränzle, W. Haber and W. Schröder describe the German experiences of environmental monitoring and introduce these into the international discussion on harmonization of global environmental monitoring and the establishment of a GTOS. As in Germany, in developing any practical environmental monitoring system at the global scale it is necessary to include existing measurement networks and to take account of the complex national and other areas of authority and jurisdiction. Both ecosystem theory (top down approach) and quality assurance and control (bottom up approach) are of decisive importance for the harmonized use of existing national and international environmental monitoring systems.

A selection of existing environmental organizations, programmes and networks are described in Annex II. This is in part a summary of some of the background information distributed to the workshop participants. The two other global observing systems, GCOS and GOOS, are included with more detail information than in the main chapters, together with some terrestrial observing systems of global or regional relevance, some of which are referred to elsewhere in this volume.

Annexes

Annex I
Participants in the Workshop on Harmonization of Environmental Data, 7–10 June 1994, Munich, Federal Republic of Germany

Prof. Dr. Marion F. Baumgardner
Purdue University
Department of Agronomy
West Laffayette, IN 47907-1150
U.S.A.
Tel: +1 317 494 5115
Fax: +1 317 496 1368
e-mail: mbaumgardner@dept.agry.purdue.edu

Dr. Inga Bucher-Wallin
Swiss Federal Institute for Forest,
Snow and Landscape Research
Zürcherstr. 111
CH-Birmensdorf
SWITZERLAND
Tel: +41 1 739 2486
Fax: +41 1 739 2488
e-mail: inga.bucher@wsl.ch
e-mail GTOS: gtos@wsl.ch

Dr. Marion E. Cheatle
Programme Officer
UNEP/GEMS-PAC
P.O. Box 30552
Nairobi
KENYA
Tel: +254 2 62 35 20
Fax: +254 2 22 64 91
e-mail: gems.unep@un.org

Mr. Josef Cohen
Head Land & Information Unit
Division of Science and Management
Nature Reserves Authority
78 Yirmeyahu St.
Jerusalem 94467
ISRAEL
Tel: +972 2 38 7471
Fax: +972 2 38 3405

Dr. Ian K. Crain
The Orbis Institute
P.O. Box 20185
Ottawa, Ontario K1N 9P4
CANADA
Tel.: +1 613 744-5653
Fax: +1 613 725 0643
e-mail: crain@arc.ab.ca

Dr. Arthur L. Dahl
Coordinator, UN System-Wide Earthwatch
EARTHWATCH Secretariat
UNEP
P.O.Box 356
CH 1219 Chatelaine Geneva
SWITZERLAND
Tel: +41 22 979 9207/9111
Fax: +41 22 797 3471
e-mail: adahl@unep.ch

Dir. Prof. Dr. Ulrich Dämmgen
Institut für agrarrelevante Klimaforschung
der FAL
Eberswalderstr. 84F
D-15374 Müncheberg
GERMANY
Tel: +49 33432 89241/42
Fax: +49 334 32 89243

Dr. Gisbert Glaser (corresponding)
Director, Bureau for Coordination of
Environmental Programme
UNESCO
7, Place de Fontenoy
Paris
FRANCE
Tel: +33 1 4568 1000
Fax: +33 1 4065 9897

Prof. Dr. Dr. h.c. W. Haber
Technische Universität München
Lehrstuhl für Landschaftsökologie
Hohenbacherstr. 19
D-8050 Freising-Weihenstephan
GERMANY
Tel: +49 8161 713 495
Fax: +49 8161 714 427

Dr. Gisela Helbig
Bundesministerium für Forschung
und Technologie
Ref. 521
Heinemannstr. 2
D-53175 Bonn
GERMANY
Tel: +49 228 593 568
Fax: +49 228 593 601

Mr. H. Franz Kerner
TEAM 18 – Umweltforschung
Bahnhofstr. 18
D-85354 Freising
GERMANY
Tel: +49 8161 530112
Fax: +49 8161 41384

Dr. Gerd Henning Klein
Projektträger Umweltsystemforschung
Südstr. 125
D-53175 Bonn
GERMANY
Tel: +49 228 3821 176
Fax: +49 228 3821 256

Dr. Hartmut Keune
Director, UNEP-HEM
c/o GSF Neuherberg
P.O. Box 1129
D-85758 Oberschleißheim
GERMANY
Tel: +49 89 3187 4418
Fax: +49 89 3187 3325
e-mail: keune@gsf.de

Dir. Prof. Dieter P. Kohnke
Bundesamt fuer Seeschiffahrt
und Hydrographie (BSH)
Bernhard-Nocht-Str. 78
D-20359 Hamburg
GERMANY
Tel: +49 40 3190 3400
Fax: +49 40 3190 5000
e-mail: kohnke@itd-
c.m4.hamburg.bsh.d400.de
omnet: d.kohnke

Ms. A. Mandy Lane
ECN Data Manager
Institute of Terrrestrial Ecology
Merlewood Research Station
Windermere Road, Grange-over-Sandes
Cumbria LA11 6JU
UNITED KINGDOM
Tel: +44 5395 32264
Fax: +44 5395 34705
e-mail: m.lane@ite.ac.uk

Dr. Roman J.M. Lenz
GSF Projekt Umweltgefährdungspotentiale
von Chemikalien
Neuherberg
Postfach 1129
D-85758 Oberschleißheim
GERMANY
Tel: +49 89 3187 2954
Fax: +49 89 3187 3369
e-mail: lenz@gsf.de

Dr. Patricia Mandry
UNEP-HEM
c/o GSF Neuherberg
P.O. Box 1129
D-85758 Oberschleißheim
GERMANY
Tel: +49 89 3187 4418
Fax: +49 89 3187 3325

Ms. Julia Marton-Lefréve (corresponding)
Executive Director
ICSU
5, Boulevard de Montmorency
F-75016 Paris
FRANCE
Tel: +33 1 4525 0329
Fax: +33 1 4288 9431

Mr. Jim McKenna
UNEP-HEM
c/o GSF Neuherberg
P.O. Box 1129
85758 Oberschleißheim
GERMANY
Tel: +49 89 3187 3364
Fax: +49 89 3187 3325
e-mail: mckenna@gsf.de

Prof. Dr. Volker A. Mohnen (corresponding)
State University of New York
University at Albany
c/o ASRC
100 Fuller Road
Albany NY 12205
U.S.A.
Tel: +1 518 442 3819
Fax: +1 518 442 3867

Dr. A. Beatrice Murray
UNEP-HEM
c/o GSF Neuherberg
P.O. Box 1129
D-85758 Oberschleißheim
GERMANY
Tel: +49 89 3187 4418
Fax: +49 89 3187 3325

Dr. Hamid Narjisse
Department of Range Science
Utah State University
Logan, Utah 84322-5230
U.S.A.
Tel: +1 801 797 1587
Fax: +1 801 797 3796
e-mail: fahamid@cc.usu.edu
(beginning January 1995)
IAV. Hassan II
BP. 6202
RABAT. Morocco
Tel: 7+212 7 7770 18
Fax: 7+212 7 7758 38

Dr. Jürgen Nauber
MAB-Geschäftsstelle
c/o BfN
Konstantinstr. 110
D-53179 Bonn
GERMANY
Tel: +49 228 8491 138
Fax: +49 228 8491 200

Prof. Henry Nix
Director, Centre for Resource
and Environmental Studies
The Australian National University
Canberra ACT 0200
AUSTRALIA
Tel: +61 62 494588
Fax: +61 62 471037 / 490757
e-mail: nix@cres.anu.edu.au

Dr. David Norse
Chairman GTOS Planning Group
ODI
Regent's College
Inner Circle
Regent's Park
London NW1 4NS
U.K.
Tel: +44 71 487 7413
Fax: +44 71 487 7590

Prof. L.A. Ogallo
Secretary
National Council for Science and Technology
Emperor Plaza, 2nd floor
P.O.Box 30623
Nairobi
KENYA
Tel: +254 2 337628/ 336173-6
Fax: +254 2 330947/ 567888-89

Dr. Kathrin Peter
Schweizerische Kommission
für Umweltbeobachtung
Projektleitung und Koordinationsstelle
Bärenplatz 2
CH-3011 Bern
SWITZERLAND
Tel: +41 31 312 3336
Fax: +41 31 312 3291

Prof. Yuri A. Pykh
President, INENCO
Russian Academy of Sciences
4 Chernomorsky per.
St. Petersburg 190000
RUSSIA
Tel: +7 812 311 76 22
Fax: +7 812 311 85 23
e-mail: pykh@inenco.spb.su

Prof. Dr. F. J. Radermacher
Forschungsinstitut für anwendungs-
orientierte Wissensverabeitung
Universität Ulm
P.O.Box 2060
D-89010 Ulm
GERMANY
Tel: +49 731 501 101
Fax: +49 731 501 999
e-mail: egner@faw9370.faw.uni-ulm.de

Mr. Walter Radermacher
Statistisches Bundesamt
Abteilung IV
D-65180 Wiesbaden
GERMANY
Tel: +49 611 752 223
Fax: +49 611 724 000/001

Mr. Lars-Otto Reiersen
General Secretary AMAP
P.O.Box 8100 Dep.
N-0032 Oslo
NORWAY
Tel: +47 22 573544/573400
Fax: +47 22 676706

Dr. Pál Ribaries
Max-Planck-Institut für Physik
Föhringer Ring 6
80805 München
GERMANY
Tel: +49 89 32 30 8219

Dr. W.-F. Riekert
Forschungsinstitut für anwendungs-
orientierte Wissensverabeitung
Universität Ulm
P.O.Box 2060
D-89010 Ulm
GERMANY
Tel: +49 731 501 101
Fax: +49 731 501 999

Dr. E. Fred Roots
Science Advisor Emeritus
Department of the Environment
K1A OH3 Ottawa, Ontario
CANADA
Tel: +1 819 997 2393
Fax: +1 819 997 5813

Dr. Winfried Schröder
Projektzentrum „Ökosystemforschung
Bornhöveder Seenkette"
Geographisches Institut der
Christian-Albrechts-Universität
Schauenburger 112
D-24098 Kiel
GERMANY
Tel: +49 431 880 3430
Fax: +49 431 880 4658
e-mail: gustav@pz-oekosys.uni-kiel.d400

Mr. Michael Sittard
ESRI
Ringstr. 7
D-85402 Kranzberg
GERMANY
Tel: +49 8166 380
Fax: +49 8166 3061

Mr. W.G. Sombroek (corresponding)
Director, Land and Water
Development Division
FAO
Viale delle Terme di Caracalla
I-00100 Rome
ITALY
Tel: +39 6 5225 3964
Fax: +39 6 5225 6275

Dr. Tom W. Spence (corresponding)
Director, GCOS
Joint Planning Office
c/o WMO
P.O.Box 2300
CH-1211 Geneva 2
SWITZERLAND
Tel: +41 22 730 8401
Fax: +41 22 740 1439

Dr. David Stanners (corresponding)
Comission European Communities, DGXI
European Environment Agency
Rue de la Loi 200
B-1049 Brussels
BELGIUM
Tel: +32 2 296 8810
Fax: +32 2 296 9562

Prof. G. Bruce Wiersma
Dean, College of Natural Resources,
Forestry & Agriculture
Maine Agricultural and Forest Experiment Station
University of Maine
Room 105
5782 Winslow Hall
Orono, Maine 04469-5782
U.S.A.
Tel: +1 207 581 3202
Fax. +1 207 581 3207
e-mail: wiersma@maine.maine.edu

Annex II

1   Global Climate Observing System (GCOS)

Introduction and objectives of GCOS

The prospect of global climate change has recently become a major concern. As increasing concentration of greenhouse gases provide the potential for climate modification unprecedented in man's history, climate issues have moved to the forefront of the international political agenda. It is essential that scientific and technical attention be focused on documenting the present state of the earth's climate, monitoring its condition, and developing an understanding of its evolution.

GCOS, the Global Climate Observing System, was established to provide the observations needed to meet the scientific requirements for monitoring the climate, detecting climate change, and for predicting climate variations and change. These observations should reduce the uncertainties regarding climate change, and help to improve models which are being developed for predicting climate variability and change.

In planning for GCOS, it was recognized that an integrated view of the requirements for all the climate system components – the global atmosphere, the world oceans, land surface, cryosphere, biosphere – must be taken.

The GCOS programme

GCOS was established by a Memorandum of Understanding (MOU) among the World Meteorological Organization, the Intergovernmental Oceanographic Commission of UNESCO, the United Nations Environment Programme, and the International Council of Scientific Unions. The MOU established a Joint Scientific and Technical Committee (JSTC) and a Joint Planning Office (JPO) to develop the plans and strategy for implementation of the system.

The first priority in the GCOS strategy developed by the JSTC is to define and develop an Initial Operational System (IOS) for observations and data. The IOS should consist of: 1) those essential observational components which are currently operational, 2) those necessary enhancements which may be specified and implemented, and 3) a comprehensive data system.

The JSTC has established a number of disciplinary oriented panels and cross cutting panels to assist in its work. In particular, a Data Management Task Group met in March 1994 to develop an initial plan for GCOS Data Management. The plan is circulating to members of the Task Group for comment, and will presented to the JSTC at the next meeting in September.

The data system will specify procedures and guidelines for collection, quality control, comparison of observations from different sources, dissemination and utilization of all data relevant to GCOS. (Determination of relevant data is the responsibility of the various disciplinary panels and scientific expertise being consulted on behalf of GCOS.)

The IOS is expected to evolve over the next decade: some elements are operational now, some enhancements may be articulated and could be provided now, and still others are components of future research and development programmes. The data management components are expected to evolve similarly.

Data management objectives

The crucial role of data and information systems and the importance of data management in support of global climate change research are now clearly recognized. The wide variety of types and sources of global, regional and local data needed for global climate studies means that existing climate data management programmes must be strengthened and extended to document what data exists, assess and improve its quality and accessibility, and foster data exchange.

The principal objective of the GCOS data management effort is to ensure that critical data and information requirements of its users/participants are satisfied. Three categories of users/participants are envisioned: researchers, operational users, and assessor/integrators. In order to satisfy the requirements of these groups most effectively, GCOS will adopt a two-level approach for its data system. The primary focus of the system will be on the needs of the research and operations users (i.e. technical participants). The needs of the specialised assessors/integrators are dependent upon the output of the data system but not specifically catered for by the system.

To meet the requirements of all of its users, GCOS must provide for the effective acquisition, processing, storage, cataloguing, documentation, retrieval, and dissemination of data from a wide range of scientific disciplines, from ground-based and space-based sources, and spanning a variety of time and space scales. Furthermore, it must ensure that all of these activities are carried out for the long periods and in the consistent manner needed to meet the exacting requirements of global climate system research.

Data system vision and strategy

At the meeting of the Task Group, a vision for the twenty-first century was discussed since GCOS is a very long-term programme and must take a long-range view of requirements for its data and information system. Planning for data access desires and capabilities ten or more years hence requires both a strategic vision of the system desired and a prediction of technological advances that are possible and likely over the next decade. Based on that information GCOS should design a system that will make best use of the anticipated advances while being flexible enough to respond to unanticipated developments. Acknowledging that long-term forecasts of technological progress are always uncertain, the following advances now seem probable within a decade.

a) Every likely participant will have access to desk-top computers with graphics capabilities. In developed countries these personal workstations will also support digital video, digital audio, and teleconferencing.

b) National and international telecommunications networks will expand significantly. National networks will provide gigabit/second rates for researchers while megabits/sec should be common between developed countries. Networks within and between most developing countries will support affordable rates on the order of 64 kilobits/sec.

c) With the advent of a large market for digital television, storage technology will advance significantly. Desk top computers will commonly have local access to many gigabytes of on-line data while large data centres will be capable of holding thousands of terabytes on-line.

Given the projected technology, what are users likely to require and expect from a data system in the early years of the next century? To illustrate what is possible, the Task Group considered several scenarios – characterising what might be possible in a relatively advanced countries or in less developed countries. It then addressed what would be required to make these scenarios possible and noted that the most important elements of the data system are:

- *International communication networks.* Networks must be capable of carrying binary data, have bandwidth sufficient for real-time interaction with data centres located potentially anywhere in the world, and be interconnected so information can pass freely on them.

- *Advanced quality control and validation systems.* Systems must ensure that the huge volumes of data collected meet the demanding requirements of global change research.

- *Production of assimilated and special data products.* To realize the full potential of environmental observations, data from different sources must be processed and combined, often through conversion to gridded fields which, to ensure consistency, may be best performed via four-dimensional data assimilation schemes run in conjunction with numerical models.

- *International standards for exchange of retrospective data.* To use environmental data efficiently, it must be provided in a readily-usable form, be delivered in at least one of possibly many standard formats, and be accompanied by comprehensive metadata that describe the data sufficiently to allow automated processing.

- *An integrated international database.* To locate and utilize the required data, it must be described in and be accessible from advanced data systems. These systems should be connected and operate such that data stored at different sites are accessible as if stored in a single location. They must provide functions to locate data, to determine if what is needed is available, to examine supporting information and metadata, and to order data to be delivered via a variety of means. High-use data sets or data that are required in real time should be available on-line; other data should be available form rapid retrieval systems. For participants without on-line access, high-level information should be collected and distributed on inexpensive media.

- *Advanced scientific workstations.* With the huge increase in data volume that is anticipated within ten years, it will not be sufficient to present data as tables of numbers. Instead, investigators will need a comprehensive set of tools to manipulate, analyze and visualize climate data. These tools should minimize the effort required to load and convert data into a format that local software can read. Participating states and institutions must be encouraged to share software and to develop software with the needs of other GCOS participants.

Strategy

If the GCOS system is to be realized, each of the capabilities described above should be developed. However, the GCOS data system will not be implemented by a central authority. Instead, it will rely upon existing institutions and countries, each with finite resources, to develop components of the system. These components must be developed in a manner that ensure their compatibility. The implementation of an integrated system through a loose international confederation will be an exceptional challenge.

In order to carry out the required activities outlined in this plan within the specified constraints, the GCOS data management strategy depends upon the following guiding principles:

a) Make maximum use of existing programmes and expertise;
b) Use international standards;
c) Set requirements of participants as a primary concern;
d) Ensure data are of appropriate quality and consistency;
e) Monitor and evaluated the system constantly.

Key responsibilities.

The JSTC is responsible for formulating the overall concept and scope of GCOS and providing scientific and technical guidance. It will develop necessary plans and stimulated (and oversee as necessary) the implementation of those plans through appropriate national and international agencies.

The Data Management Task Group recommended a Data Management Panel reporting to the JSTC, to ensure that a comprehensive data system is available to meet GCOS requirements by coordinating and overseeing development and implementation of this plan.

It must be emphasized that the implementation and technical management of the GCOS data system will be through existing national and international agencies and organizations. The data system will be developed, implemented and operated by these organizations and programmes and the final success of the programme rests in their hands. As additional resources may be needed for these organizations to meet the additional responsibilities that GCOS imposes, GCOS will strive to ensure that their requirements for support are considered by the appropriate national and multi-national funding agencies.

Many components of an international data system are already in place and a modest effort to standardize interactions and build bridges between systems can provide a substantial return. Many countries operate national centres which maintain local climatological, hydrological, and natural resource data. The global observation, communication and processing systems for operational real-time meteorological and marine data are existing responsibilities of the World Wheather Watch (WWW). The WWW Global Data Processing Centres and ICSU World Data Centres have extensive capabilities for processing, validating, cataloguing, retrieving, and disseminating environmental data.

As GCOS will be built upon the existing and planned international data management infrastructure, it is essential that it determine the full extent of these activities and evaluate the role they should play in the development of the GCOS data system. Clearly such assessment must be done in concert with other national and international agency efforts.

Conclusion

The role of GCOS is to define the requirements for climate observations, and with the participation of national and international agencies and organization, effectively utilize existing structures to the extent possible to obtain the necessary observations and develop an effective data management system. Where capabilities do not exist, it will be important for GCOS to disclose deficiencies, and to work cooperatively to redress them.

This paper was submitted to UNEP-HEM for the workshop. Additional information can be obtained from: Thomas W. Spence, Global Climate Observing System, Joint Planning Office c/o Meteorological Organization, P.O.Box 2300, CH 1211 Geneva 2, Switzerland; Tel: +41 22 730 8401; Fax: +41 22 740 1439.

2 Global Ocean Observing System (GOOS)

Introduction

The Global Ocean Observing System (GOOS) has been initiated by the Intergovernmental Oceanographic Commission (IOC), in cooperation with the International Council of Scientific Unions (ICSU), the World Meteorological Organization (WMO), and the United Nations Environment Programme (UNEP). GOOS is supported by the Second World Climate Conference (SWCC) and by the United Nations Conference on Environment and Development (UNCED). GOOS is being developed in a phased approach. The first phase, 1990–2000, associated with the implementation of large-scale research programmes, will involve the formulation of the overall policy, the scientific, technological, and administrative and management basis, the refinement of general plans and establishment of national and international infrastructures needed for GOOS. It is anticipated that a fully operational GOOS can be established by about the year 2007.

Objectives and scope

The objective of GOOS is to ensure global, permanent, systematic observations adequate for forecasting climate variability and change, for assessing the health or state of the marine environment and its resources, including the coastal zone; and for supporting an improved decision-making and management process, which takes into account their effects on human health and resources.

GOOS is an internationally coordinated system for systematic operational data collection (measurements), data analysis, exchange of data and information, and technology development and transfer. GOOS will use a globally-coordinated, scientifically-based strategy to allow for monitoring and subsequent prediction of environmental changes globally, regionally and nationally.

Emphasis will be placed on the open exchange of data with data bases accessible to all participating countries. Individual efforts will be very much enhanced by access to a much larger cooperative data set. Because of the magnitude of the endeavour and the potential benefits, all countries are encouraged to participate.

Basic approach

GOOS will be established by Member States and implemented through nationally-owned and operated facilities and services. Coordination is provided by IOC in cooperation with WMO; UNEP and ICSU. GOOS is based on the principle that all countries should participate and that participants should make certain commitments, according to their capabilities, so that all countries can benefit.
GOOS will be developed on a sound scientific basis using the findings of existing, on-going research programmes including the World Ocean Circulation Experiment (WOCE) programme, the Tropical Oceans and Global Atmosphere (TOGA) programme, the Joint Global Ocean Flux Studies (JGOFS) and the Land Ocean Interaction in the Coastal Zone (LOICZ) programme. Operational programmes include the Integrated Global Ocean Services System (IGOSS), the International Oceanographic Data and Information Exchange (IODE), the Global Sea Level

Observing System (GLOSS) and the Data Buoy Cooperation Panel (DBCP). Observations should be: i) long-term; ii) systematic; iii) relevant to the global system; iv) cost-effective; and v) routine. GOOS will utilize a variety of operational observing methods, including both remote sensing and *in-situ* measurements. GOOS will promote the enhancement of these systems and will consider use of newly-designed systems such as powered sub-surface vehicles and long-range acoustic propagation through the ocean.

The major elements of GOOS are operational, oceanographic observations and analyses, timely distribution of data and products, data assimilation into numerical models leading to predictions, and capacity building within participating Member States, especially in developing countries to develop analysis and application capability.

GOOS will not only meet climate-related needs. Five modules have been defined at present which represent a range of interests and applications: (i) Climate Monitoring, Assessment and Prediction; this module is common with the ocean component of GCOS – the Global Climate Observing System; (ii) Monitoring and Assessment of Marine Living Resources; (iii) Monitoring of the Coastal Zone Environment and its Changes; (iv) Assessment and Prediction of the Health of the Ocean; and (v) Marine Meteorological and Oceanographic Operational Services. These modules are inter-related and will share observations, data and data networks and facilities, as needed, within the one integrated system, resulting in a synergistic effect.

GOOS will be developed in a phased approach: (i) a planning phase including conceptualization, design and technical definition; (ii) operational demonstrations for each of the five modules; (iii) implementation of permanent aspects of the GOOS; and (iv) continued assessment and improvement in the individual aspects of the entire system.

Conclusion

Today we are experiencing unprecedented pressures on our natural resources. Sustainable development of these resources is hindered by our inability to detect emerging environmental problems at an early stage when remedial measures are still possible. Nowhere is this inadequacy so pronounced as in the marine area. Global energy cycles and the biological processes upon which all life depends are critically influenced by the ocean. Governments collectively are only now beginning to recognize the complexity and interdependence of all aspects of the system. Systematic global observations of the world oceans are required to improve our knowledge and predictive capabilities which will be the basis for more effective and sustained use of the marine environment, with the associated economic benefits. GOOS is intended to fulfil this need.

This information is extracted from a paper submitted to UNEP-HEM for the workshop. Additional information may be obtained from: IOC Secretariat, UNESCO GOOS Support Office 1, rue Miollis 75732 Paris Cedex 15, France; Tel: +33 1 45684042; Fax: +33 1 40569316; Telemail: GOOS.PARIS/OMNET.

3 Terrestrial Observing Systems of Global or Regional Relevance

3.1 EuroMAB Biosphere Reserve Integrated Monitoring (BRIM) for Biological Diversity Conservation and Sustainable Use of Natural Resources

Background

The European Man and Biosphere (MAB) National Committees have founded a sub-network called EUROMAB. Every two years they meet in conference in order to evaluate the results of their work and to discuss the further development of their cooperation. In 1991, at the Third Conference of the European and North American MAB National Committees (EUROMAB III) in Strasbourg, the delegates recognised the need to implement appropriate international cooperation to utilise the potential of the UNESCO biosphere reserves, and established a harmonized integrated monitoring programme with the name "Biosphere Reserves Integrated Monitoring Programme" (BRIM). Overall objective is to give access to the potential for monitoring and research of the 180 Biosphere Reserves in Europe.

The BRIM-programme will take advantage of the networking function of the MAB programme and the Biosphere Reserves in order to contribute to global environmental monitoring. Biosphere Reserves are integrated horizontally and vertically. The horizontal integration unites international high level interdisciplinary science and monitoring. By the vertical integration the individual Biosphere Reserve on the ground itself is directly connected via its government with international agencies and the international scientific community.

BRIM should be seen as a pilot programme and it is expected that the other regional networks of MAB will join the initiative. The contribution of MAB to GTOS is one of the five priorities of the future MAB programme as the 12th MAB International Coordinating Council has decided at its meeting in January 1993.

BRIM's vision

A harmonized network for integrated ecological monitoring initially centered on European and North American Biosphere Reserves of the Man and the Biosphere (MAB) Program.

BRIM's mission:

- To organise and share information from the biological, physical and social science disciplines that is generated by research and monitoring in biosphere reserves within common biomes;

- To improve the reliability of information concerning international environmental trends;

- To provide a substantial contribution to research and monitoring for sustainable use and conservation of biological diversity.

Goals

- To provide access for the scientific, administrative and policy communities to the biological, physical and social science information data bases that are available on the biosphere reserves of Europe and North America.

- To provide a means for a systematic exchange of scientific information and the integrated monitoring of biosphere reserves, with special emphasis on global change, biological diversity, ecosystems management and human impact and environmental sustainability.

Implementing BRIM by organising networks of monitoring sites

First:
Establish a broad based system of comparable data sets on the maximum number of EUROMAB Biosphere Reserves.

BRIM focuses its priorities on:

1. *Adoption of Standardized Protocols for reporting inventories of flora and fauna.* BRIM protocols and data bases will help identify: species rich sites; unique flora and fauna; and status of species information among EuroMAB Biosphere Reserves.

2. *Establish Permanent Plots for Monitoring and Research.* BRIM will produce a directory on permanent plots (existing and planned) on EUROMAB Biosphere Reserves. The directory will offer guidelines and criteria for permanent plots.

3. *Complement the World Meteorological Organization (WMO) network sites for climatological monitoring.*

Second:
To develop in-depth research and monitoring networks among EUROMAB Biosphere Reserves which have longer historical records, more complete information, specialised research and monitoring activities, or strong current research and monitoring activities in the fields of : effects of human activities, vertebrates, invertebrates, plants and precipitation chemistry.

Third:
In collaboration with the UNESCO MAB Secretariat, work to establish a central clearing house and database for biosphere reserve information.

Priority efforts for BRIM/UNESCO/MAB initially focuses on:

1. Promoting communication between all biosphere reserves;

2. Encouraging standardized data and information reporting formats in priority areas

3. Providing standards for quality control over data.

4. Facilitating communication between the world's scientific community and biosphere reserves; promoting use of EUROMAB data in monitoring global climate change, social trends and initiatives, and promoting sustainable use of natural resources and the implementation of the Convention on biological diversity.

Fourth:
Establish guidelines and criteria for monitoring the effects of human activities in biosphere reserves.

BRIM strategy and on-going activities

As the first result of the international cooperation, a directory of the European Biosphere Reserves, ACCESS, was published by the MAB National Committee of the USA. It gives all the contact addresses for the approximately 180 European Biosphere Reserves and informs about 70 areas of activities each Biosphere Reserve is working in (for summary information see Table 1). The UNESCO/MAB Secretariat has recently published a similar directory for the 320 Biosphere Reserves world-wide.

The second step is the creation of a database which gives information about the potential of permanent plots of all kinds in the Biosphere Reserves for monitoring and research. These are important because monitoring and research usually take place on permanent plots rather than on large areas. The German MAB National Committee is currently conducting an inquiry on the permanent plots of the European Biosphere Reserves. The results will yield information about the potential of the Biosphere Reserves for monitoring and observation purposes. It will also indicate how the European ecosystem types are represented by the Biosphere Reserve Network. The results of the inquiry will be published as ACCESS II probably in spring 1995.

Figure 1 shows the number of European Biosphere Reserves which have monitoring activities in the respective thematic fields.

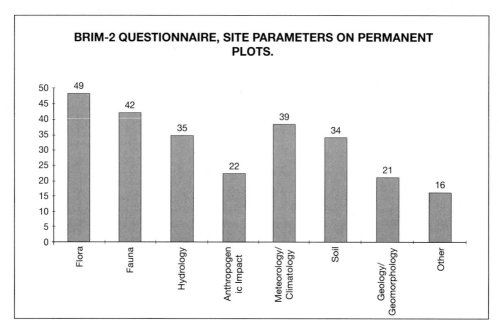

Fig. 1:
BRIM Questionnaire. Site Parameters on Permanent Plots

Figure 2 summarises as an example the number of Biosphere Reserves which yield data on specific themes on flora dedicated permanent plots.

The third step of BRIM is the development of a standard instrument for inventorying flora and fauna in the Biosphere Reserves. The U.S.A. MAB-National Committee in cooperation with the US-National Park Service has elaborated a computer programme for the inventory of flora and fauna to be used by all 180 European Biosphere Reserves. The programme will be distributed to all European Biosphere Reserves in the course of 1994 and 1995.

The fourth step is the design of special monitoring and research projects which are concentrated on the one hand on special MAB issues, on the other on global monitoring issues for which the Biosphere Reserves are suitable sites as seen from their ecological composition.

It is planned that all results of the BRIM-programme will be available to interested parties through INTERNET.

This information is from a paper submitted to UNEP-HEM for the workshop. Additional information on BRIM may be obtained from: Jürgen Nauber, MAB Geschäftsstelle c/o BfN, Konstantinstr. 110, D-53179 Bonn, Germany; and Dr. Roger E. Soles, Executive Director, U.S MAB Secretariat, OES/EGC/MAB, Room 608 SA 37, Department of State, Washington, D.C. 20522-3707, U.S.A.

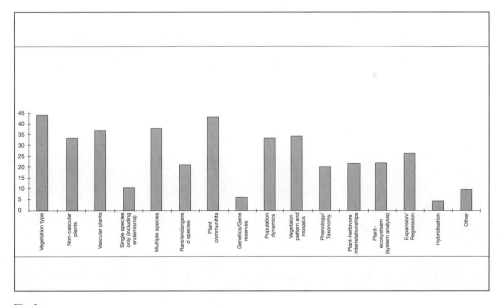

Fig. 2:
Data on Flora Dedicated to Permanent Plots

Table 1:
EUROMAB biomes and states

	TEMPERATE BROAD-LEAF FOREST	COMPLEX MOUNTAIN SYSTEMS	EVERGREEN SCLEROPHYLLOUS WOODLAND	TEMPERATE NEEDLE-LEAF FOREST	TEMPERATE GRASSLAND	MIXED ISLAND SYSTEMS	WARM DESERT	TUNDRA	TEMPERATE RAINFOREST	COLD DESERT	LAKE SYSTEMS	TROPICAL DRY FOREST	COUNTRY TOTAL	COUNTRY % OF TOTAL
AUSTRIA	1	3	0	0	0	0	0	0	0	0	0	0	4	2%
BELARUS	1	0	0	0	0	0	0	0	0	0	0	0	1	1%
BULGARIA	1	16	0	0	0	0	0	0	0	0	0	0	17	10%
CANADA	2	1	0	2	0	0	0	0	0	0	1	0	6	3%
CROATIA	0	0	1	0	0	0	0	0	0	0	0	0	1	1%
CZECH REPUBLIC	2	2	0	0	0	0	0	0	0	0	0	0	4	2%
CZECH. & POLAND	0	1	0	0	0	0	0	0	0	0	0	0	1	1%
DENMARK	0	0	0	0	0	0	1	0	0	0	0	0	1	1%
ESTONIA	1	0	0	0	0	0	0	0	0	0	0	0	1	1%
FINLAND	0	0	0	1	0	0	0	0	0	0	0	0	1	1%
FRANCE	3	0	3	0	0	2	0	0	0	0	0	0	8	5%
GERMANY	10	2	0	0	0	0	0	0	0	0	0	0	12	7%
GREECE	0	0	2	0	0	0	0	0	0	0	0	0	2	1%
HUNGARY	5	0	0	0	0	0	0	0	0	0	0	0	5	3%
IRELAND	2	0	0	0	0	0	0	0	0	0	0	0	2	1%
ITALY	0	1	2	0	0	0	0	0	0	0	0	0	3	2%
KYRG. & UZBEK.	0	1	0	0	0	0	0	0	0	0	0	0	1	1%
NETHERLANDS	1	0	0	0	0	0	0	0	0	0	0	0	1	1%
NORWAY	0	0	0	0	0	0	0	1	0	0	0	0	1	1%
POLAND	4	0	0	0	0	0	0	0	0	0	0	0	4	2%
POLAND & SLOVAKIA	2	0	0	0	0	0	0	0	0	0	0	0	2	1%
PORTUGAL	0	0	1	0	0	0	0	0	0	0	0	0	1	1%
ROMANIA	2	0	0	0	1	0	0	0	0	0	0	0	3	2%
RUSSIAN FEDERATION	6	2	0	4	1	0	0	0	0	1	0	0	14	8%
SLOVAKIA	2	0	0	0	0	0	0	0	0	0	0	0	2	1%
SPAIN	0	0	10	0	0	1	0	0	0	0	0	0	11	6%
SWEDEN	0	0	0	0	0	0	0	1	0	0	0	0	1	1%
SWITZERLAND	0	1	0	0	0	0	0	0	0	0	0	0	1	1%
TURKMENISTAN	0	0	0	0	0	0	0	0	0	1	0	0	1	1%
UKRAINE	1	0	0	0	2	0	0	0	0	0	0	0	3	2%
UNITED KINGDOM	6	6	0	0	0	0	0	0	0	0	0	0	12	7%
UNITED STATES	11	10	4	1	3	4	5	2	4	1	1	1	47	27%
YUGOSLAVIA	0	1	0	0	0	0	0	0	0	0	0	0	1	1%
BIOME TOTAL	63	47	23	8	7	7	5	5	4	3	2	1	175	
% OF EUROMAB BR	36%	27%	13%	5%	4%	4%	3%	3%	2%	2%	1%	1%		

Table 2:
EUROMAB biomes and basic resource information

BASIC RESOURCE INFORMATION
NUMBER OF BIOSPHERE RESERVES IN BIOME REPORTING INFORMATION FOR CATEGORY

BIOME:	BIOLOGICAL INVENTORY							ECOLOGICAL MONITORING										RESOURCE MAPS					
	INVERTEBRATES	MAMMALS	BIRDS	NONVASCULAR PLANTS	VASCULAR PLANTS	VERTEBRATES OTHER THAN MAMMALS	BIOLOGICAL SURVEY AND COLLECTIONS	AIR QUALITY	CLIMATE	FRESHWATER ECOSYSTEMS	GROUNDWATER HYDROLOGY	MARINE ECOSYSTEMS	PALEOECOLOGY	PRECIPITATION CHEMISTRY	SURFACE HYDROLOGY	VEGETATION DATA	WATER QUALITY	GEOLOGICAL	LAND USE	REGIONAL LAND TENURE	SOILS	TOPOGRAPHIC	VEGETATION
TEMPERATE RAINFOREST	3	4	2	3	4	3	4	2	4	4	2	3	2	1	3	4	2	4	3	2	2	3	3
TEMPERATE NEEDLE-LEAF FOREST	5	8	3	5	8	7	7	5	6	4	2	1	1	4	6	4	5	5	3	2	5	5	7
TROPICAL DRY FOREST	0	1	1	0	1	1	1	1	1	1	1	1	0	1	1	1	1	1	0	0	1	1	1
TEMPERATE BROAD-LEAF FOREST	56	60	22	50	57	56	57	35	57	42	40	13	11	30	52	42	49	48	38	13	49	55	56
EVERGREEN SCLEROPHYLLOUS WOODLAND	16	18	11	17	20	20	18	8	20	12	8	5	2	6	16	19	15	20	18	7	19	20	17
WARM DESERT	4	4	4	3	4	4	4	2	5	3	4	1	2	3	5	4	4	4	2	2	5	5	5
COLD DESERT	1	2	1	1	2	2	3	2	3	1	2	0	1	0	2	3	0	2	1	1	3	3	3
TUNDRA	2	4	0	1	4	4	4	1	4	3	4	2	0	2	2	4	1	3	2	1	2	2	4
TEMPERATE GRASSLAND	6	7	2	5	7	6	7	1	7	4	4	2	1	3	5	6	5	5	6	0	6	6	6
COMPLEX MOUNTAIN SYSTEMS	29	33	10	19	38	39	40	16	32	18	13	0	6	16	22	37	22	39	28	5	34	40	37
MIXED ISLAND SYSTEMS	3	3	1	4	5	3	4	2	6	3	3	2	0	1	3	4	2	4	3	0	4	5	6
LAKE SYSTEMS	1	2	2	1	2	1	2	2	1	1	0	0	2	1	2	2	2	1	1	2	2	2	2
TOTAL BIOSPHERE RESERVES (EX 170)	126	146	59	109	152	146	151	80	146	96	81	29	28	68	119	130	108	136	105	35	132	148	147
PERCENT OF TOTAL	74%	86%	35%	64%	89%	86%	89%	47%	86%	56%	48%	17%	16%	40%	70%	76%	64%	80%	62%	21%	78%	87%	86%

SUMMARY INFORMATION

Table 3:
EUROMAB biomes and research topics

BIOME	Aerial Photographs	Bibliography	Bibliography - Number of References	Bibliography - Year of Last Revision	Geographic Information System	History of Scientific Study	Biogeochemical Cycles	Comparative Ecological Research	Ecological Succession	Ecosystem Modeling	Fire History/Effects	Hydrological Cycle	Sedimentation	Pests and Diseases	Rare/Endangered Species	Wildlife Population Dynamics	Acidic Deposition	Atmospheric Pollutants	Pesticides	Water Pollutants
TEMPERATE RAINFOREST	4	3	2	2	2	2	2	2	4	2	3	1	4	2	4	4	4	2	0	2
TEMPERATE NEEDLE-LEAF FOREST	6	6	2	2	1	5	3	6	6	3	4	4	5	5	6	8	4	6	2	6
TROPICAL DRY FOREST	1	1	0	0	1	0	1	1	1	1	1	1	0	0	1	1	1	1	1	1
TEMPERATE BROAD-LEAF FOREST	58	51	18	19	22	48	26	44	51	28	23	32	41	29	55	51	35	36	18	40
EVERGREEN SCLEROPHYLLOUS WOODLAND	18	19	5	6	6	16	8	12	16	12	17	12	8	11	17	18	5	9	5	10
WARM DESERT	4	4	2	2	4	5	1	3	4	4	4	5	1	3	3	4	3	3	3	3
COLD DESERT	2	3	0	0	0	3	1	3	2	1	1	2	0	2	1	2	0	2	0	1
TUNDRA	4	3	1	1	3	4	1	3	6	2	1	1	4	1	3	4	2	3	1	3
TEMPERATE GRASSLAND	4	7	2	2	3	7	6	7	6	4	6	4	4	5	6	7	4	4	4	3
COMPLEX MOUNTAIN SYSTEMS	25	31	9	9	8	20	10	26	25	17	10	16	33	18	33	24	15	14	8	13
MIXED ISLAND SYSTEMS	5	6	1	1	3	4	3	5	5	2	2	3	3	2	5	5	1	0	1	1
LAKE SYSTEMS	2	2	2	2	1	2	1	1	2	1	1	0	1	1	2	2	1	2	0	1
TOTAL BIOSPHERE RESERVES (EX 170)	133	136	44	46	51	116	63	113	125	77	73	81	101	79	136	130	75	81	43	82
PERCENT OF TOTAL	78%	80%	26%	27%	30%	68%	37%	66%	74%	45%	43%	48%	59%	46%	80%	76%	44%	48%	25%	48%

Column groupings:
- **BASIC RESOURCE INFORMATION**: Aerial Photographs; Historical Records (Bibliography, Bibliography - Number of References, Bibliography - Year of Last Revision, Geographic Information System, History of Scientific Study)
- **RESEARCH TOPICS**:
  - *Ecosystem Cycles and Processes*: Biogeochemical Cycles, Comparative Ecological Research, Ecological Succession, Ecosystem Modeling, Fire History/Effects, Hydrological Cycle, Sedimentation
  - *Species Populations*: Pests and Diseases, Rare/Endangered Species, Wildlife Population Dynamics
  - *Pollution*: Acidic Deposition, Atmospheric Pollutants, Pesticides, Water Pollutants

Number of biosphere reserves in biome reporting information for category.

Table 3:
continued (1)

RESEARCH TOPICS

SUMMARY INFORMATION — NUMBER OF BIOSPHERE RESERVES IN BIOME REPORTING INFORMATION FOR CATEGORY

BIOME:	| HUMAN SYSTEMS |							| MANAGEMENT PRACTICES |									
	ARCHAEOLOGY	CULTURAL ANTHROPOLOGY	DEMOGRAPHY/SETTLEMENT PATTERNS	ETHNOBIOLOGY	LAND TENURE/USE-MANAGEMENT	RESOURCE ECONOMICS	TRADITIONAL LAND USE SYSTEMS	AGRICULTURAL	APPROPRIATE RURAL TECHNOLOGY	ECOSYSTEM RESTORATION	GENETIC RESOURCE MANAGEMENT	MINING RECLAMATION	RANGELAND MANAGEMENT	RECREATION/TOURISM	RESOURCE PRODUCTION TECHNOLOGIES	SOIL CONSERVATION	WATERSHED MANAGEMENT
TEMPERATE RAINFOREST	2	3	2	0	2	1	3	0	0	3	2	2	0	2	1	3	1
TEMPERATE NEEDLE-LEAF FOREST	2	5	2	2	3	1	4	1	1	4	2	5	0	4	2	5	3
TROPICAL DRY FOREST	0	0	0	0	0	0	1	0	0	1	1	1	1	1	0	1	1
TEMPERATE BROAD-LEAF FOREST	14	23	16	15	14	11	40	30	21	46	28	25	17	44	13	39	26
EVERGREEN SCLEROPHYLLOUS WOODLAND	5	12	3	6	5	4	10	10	5	16	7	6	8	17	3	13	9
WARM DESERT	2	3	1	3	1	0	3	2	2	4	0	2	2	3	1	3	3
COLD DESERT	0	0	1	0	1	1	2	0	0	2	0	1	1	1	0	1	0
TUNDRA	0	2	0	0	0	0	3	0	1	2	0	2	0	2	0	0	0
TEMPERATE GRASSLAND	2	4	1	2	4	1	5	5	4	3	3	2	5	2	2	6	4
COMPLEX MOUNTAIN SYSTEMS	6	2	3	2	3	2	17	7	3	17	9	8	9	14	5	19	7
MIXED ISLAND SYSTEMS	0	2	0	1	1	0	2	1	1	4	3	0	0	4	1	2	3
LAKE SYSTEMS	2	2	1	1	2	1	0	0	0	2	1	0	0	1	0	0	1
TOTAL BIOSPHERE RESERVES (EX 170)	35	58	30	32	36	22	90	56	38	104	56	54	43	95	29	92	58
PERCENT OF TOTAL	21%	34%	18%	19%	21%	13%	53%	33%	22%	61%	33%	32%	25%	56%	17%	54%	34%

Table 4:
EUROMAB biomes and site support

SITE SUPPORT

NUMBER OF BIOSPHERE RESERVES IN BIOME REPORTING INFORMATION FOR CATEGORY

BIOME:	INFRASTRUCTURE						MONITORING AND RESEARCH FACILITIES						
	CONFERENCE FACILITIES	CURATORIAL FACILITY	LABORATORY	LIBRARY	LODGING FOR SCIENTISTS	ROAD ACCESS	AIR POLLUTION STATION	HYDROLOGICAL STATION	PERMANENT PLOTS: LAKE/STREAM	PERMANENT PLOTS: VEGETATION	WATERSHED RESEARCH SITE	WEATHER STATION	PERMANENT RESEARCH STAFF
TEMPERATE RAINFOREST	3	2	2	3	4	2	3	2	1	3	3	4	4
TEMPERATE NEEDLE-LEAF FOREST	6	6	6	8	5	7	4	5	4	5	7	6	6
TROPICAL DRY FOREST	1	1	0	1	1	0	1	1	0	1	0	1	1
TEMPERATE BROAD-LEAF FOREST	52	31	45	45	47	46	35	38	15	39	22	53	36
EVERGREEN SCLEROPHYLLOUS WOODLAND	15	13	16	16	17	22	4	15	4	7	6	18	15
WARM DESERT	4	3	4	4	4	4	3	5	0	4	2	5	3
COLD DESERT	2	3	2	2	2	3	2	1	1	0	0	3	2
TUNDRA	2	3	2	4	1	2	2	2	0	3	1	4	3
TEMPERATE GRASSLAND	7	5	7	4	7	7	3	2	3	4	3	5	6
COMPLEX MOUNTAIN SYSTEMS	24	13	16	20	30	30	15	22	7	20	10	30	22
MIXED ISLAND SYSTEMS	4	3	4	4	4	5	0	2	3	3	3	6	5
LAKE SYSTEMS	1	1	1	2	2	1	2	1	1	1	1	2	0
TOTAL BIOSPHERE RESERVES (EX 170)	121	84	105	113	124	129	74	96	39	90	58	137	103
PERCENT OF TOTAL	71%	49%	62%	66%	73%	76%	44%	56%	23%	53%	34%	81%	61%

SUMMARY INFORMATION

Table 5a:
Frequency of databases on EUROMAB basic resource information

%	Bio. Res. #	CATEGORY
		BASIC RESOURCE INFO
		BIOLOGICAL INVENTORY
89.4%	152	VASCULAR PLANTS
88.8%	151	BIOLOGICAL SURVEY AND COLLECTIONS
85.9%	146	MAMMALS
85.9%	146	VERTEBRATES OTHER THAN MAMMALS
74.1%	126	INVERTEBRATES
64.1%	109	NONVASCULAR PLANTS
34.7%	59	BIRDS
		ECOLOGICAL MONITORING
85.9%	146	CLIMATE
76.5%	130	VEGETATION DATA
70.0%	119	SURFACE HYDROLOGY
63.5%	108	WATER QUALITY
56.5%	96	FRESHWATER ECOSYSTEMS
47.6%	81	GROUNDWATER HYDROLOGY
47.1%	80	AIR QUALITY
40.0%	68	PRECIPITATION CHEMISTRY
17.1%	29	MARINE ECOSYSTEMS
16.5%	28	PALEOECOLGY
		RESOURCE MAPS
87.1%	148	TOPOGRAPHIC
86.5%	147	VEGETATION
80.0%	136	GEOLOGICAL
77.6%	132	SOILS
61.8%	105	LAND USE
20.6%	35	REGIONAL LAND TENURE
		HISTORICAL RECORDS
80.0%	136	BIBLIOGRAPHY
78.2%	133	AERIAL PHOTOGRAPHS
68.2%	116	HISTORY OF SCIENTIFIC STUDY
30.0%	51	GEOGRAPHIC INFORMATION SYSTEM

Table 5b:
Frequency of databases on EUROMAB research topics

ECOSYSTEM CYCLES AND PROCESSES

%	#	Category
73.5%	125	ECOLOGICAL SUCCESSION
66.5%	113	COMPARITIVE ECOLOGICAL RESEARCH
59.4%	101	SEDIMENTATION
47.6%	81	HYDROLOGICAL CYCLE
45.3%	77	ECOSYSTEM MODELING
42.9%	73	FIRE HISTORY AND EFFECTS
37.1%	63	BIOGEOCHEMICAL CYCLES

%	Bio. Res.	#	CATEGORY

SPECIES POPULATIONS

%	#	Category
80.0%	136	RARE AND ENDANGERED SPECIES
76.5%	130	WILDLIFE POPULATION DYNAMICS
46.5%	79	PESTS AND DISEASES

RESEARCH TOPICS

POLLUTION

%	#	Category
48.2%	82	WATER POLLUTANTS
47.6%	81	ATMOSPHERIC POLLUTANTS
44.1%	75	ACIDIC DEPOSITION
25.3%	43	PESTICIDES

HUMAN SYSTEMS

%	#	Category
52.9%	90	TRADITIONAL LAND USE SYSTEMS
34.1%	58	CULTURAL ANTHROPOLOGY
21.2%	36	LAND TENURE/ USE/MANAGEMENT
20.6%	35	ARCHAEOLOGY
18.8%	32	ETHNOBIOLOGY
17.6%	30	DEMOGRAPHY/SETTLEMENT PATTERNS
12.9%	22	RESOURCE ECONOMICS

MANAGEMENT PRACTICES

%	#	Category
61.2%	104	ECOSYSTEM RESTORATION
55.9%	95	RECREATION AND TOURISM
54.1%	92	SOIL CONSERVATION
34.1%	58	WATERSHED MANAGEMENT
32.9%	56	AGRICULTURAL
32.9%	56	GENETIC RESOURCE MANAGEMENT
31.8%	54	MINING RECLAMATION
25.3%	43	RANGELAND MANAGEMENT
22.4%	38	APPROPRIATE RURAL TECHNOLOGY
17.1%	29	RESOURCE PRODUCTION TECHNOLOGIES

Table 5c:
Frequency of databases on EUROMAB site support

INFRASTRUCTURE

75.9%	129	ROAD ACCESS
72.9%	124	LODGING FOR SCIENTISTS
71.2%	121	CONFERENCE FACILITIES
66.5%	113	LIBRARY
61.8%	105	LABORATORY
49.4%	84	CURATORIAL FACILITY

MONITORING AND RESEARCH FACILITIES

80.6%	137	WEATHER STATION
56.5%	96	HYDROLOGICAL STATION
52.9%	90	PERMANENT PLOTS: VEGETATION
43.5%	74	AIR POLLUTION STATION
34.1%	58	WATERSHED RESEARCH SITE
22.9%	39	PERMANENT PLOTS: LAKE/STREAM
60.6%	103	PERMANENT RESEARCH STAFF

Table 5d:
Summary information: Ranking frequencies of databases

%	Bio. Res. #	CATEGORY
89.4%	152	VASCULAR PLANTS
88.8%	151	BIOLOGICAL SURVEY AND COLLECTIONS
87.1%	148	TOPOGRAPHIC
86.5%	147	VEGETATION
85.9%	146	MAMMALS
85.9%	146	VERTEBRATES OTHER THAN MAMMALS
85.9%	146	CLIMATE
80.6%	137	WEATHER STATION
80.0%	136	GEOLOGICAL
80.0%	136	BIBLIOGRAPHY
80.0%	136	RARE AND ENDANGERED SPECIES
78.2%	133	AERIAL PHOTOGRAPHS
77.6%	132	SOILS
76.5%	130	VEGETATION DATA
76.5%	130	WILDLIFE POPULATION DYNAMICS
75.9%	129	ROAD ACCESS
74.1%	126	INVERTEBRATES
73.5%	125	ECOLOGICAL SUCCESSION
72.9%	124	LODGING FOR SCIENTISTS
71.2%	121	CONFERENCE FACILITIES
70.0%	119	SURFACE HYDROLOGY
68.2%	116	HISTORY OF SCIENTIFIC STUDY
66.5%	113	COMPARITIVE ECOLOGICAL RESEARCH
66.5%	113	LIBRARY
64.1%	109	NONVASCULAR PLANTS

Table 5d:
continued (1)

63.5%	108	WATER QUALITY
61.8%	105	LAND USE
61.8%	105	LABORATORY
61.2%	104	ECOSYSTEM RESTORATION
60.6%	103	PERMANENT RESEARCH STAFF
59.4%	101	SEDIMENTATION
56.5%	96	FRESHWATER ECOSYSTEMS
56.5%	96	HYDROLOGICAL STATION
55.9%	95	RECREATION AND TOURISM
54.1%	92	SOIL CONSERVATION
52.9%	90	TRADITIONAL LAND USE SYSTEMS
52.9%	90	PERMANENT PLOTS: VEGETATION
49.4%	84	CURATORIAL FACILITY
48.2%	82	WATER POLLUTANTS
47.6%	81	GROUNDWATER HYDROLOGY
47.6%	81	HYDROLOGICAL CYCLE
47.6%	81	ATMOSPHERIC POLLUTANTS
47.1%	80	AIR QUALITY
46.5%	79	PESTS AND DISEASES
45.3%	77	ECOSYSTEM MODELING
44.1%	75	ACIDIC DEPOSITION
43.5%	74	AIR POLLUTION STATION
42.9%	73	FIRE HISTORY AND EFFECTS
40.0%	68	PRECIPITATION CHEMISTRY
37.1%	63	BIOGEOCHEMICAL CYCLES
34.7%	59	BIRDS
34.1%	58	CULTURAL ANTHROPOLOGY
34.1%	58	WATERSHED MANAGEMENT
34.1%	58	WATERSHED RESEARCH SITE
32.9%	56	AGRICULTURAL
32.9%	56	GENETIC RESOURCE MANAGEMENT
31.8%	54	MINING RECLAMATION
30.0%	51	GEOGRAPHIC INFORMATION SYSTEM
25.3%	43	PESTICIDES
25.3%	43	RANGELAND MANAGEMENT
22.9%	39	PERMANENT PLOTS: LAKE/STREAM
22.4%	38	APPROPRIATE RURAL TECHNOLOGY
21.2%	36	LAND TENURE/ USE/MANAGEMENT
20.6%	35	REGIONAL LAND TENURE
20.6%	35	ARCHAEOLOGY
18.8%	32	ETHNOBIOLOGY
17.6%	30	DEMOGRAPHY/SETTLEMENT PATTERNS
17.1%	29	MARINE ECOSYSTEMS
17.1%	29	RESOURCE PRODUCTION TECHNOLOGIES
16.5%	28	PALEOECOLGY
12.9%	22	RESOURCE ECONOMICS

3.2 Integrated Monitoring by UN-ECE

Background

Following the initiative of the Nordic countries, a Pilot Programme on Integrated Monitoring (PP/IM) was established in 1987, to be implemented under the UN ECE Convention on Long-range Transboundary Air Pollution.

It was considered that this type of monitoring would provide for a better understanding of direct and indirect effects of air pollutants and of long-term changes in the environment. Monitoring methods were agreed upon during three intensive workshops in 1988–1990, to a large extent based on the suggestions already existing in the Nordic countries. These methods were described in detail in two manuals (Field and Laboratory Manual, 1989; Manual for input to the Environment Data Centre/ Integrated Monitoring (EDC/IM) (EDC/IM Data Bank, 1989). Other countries were recommended to start monitoring and to send data to EDC. Based on the reported data preliminary evaluations have been performed in "Annual Synoptic Reports" in 1990 and 1991.

An evaluation made by leading environmental scientists of the pilot phase was performed during 1991-92 as a basis for decisions about the future of the programme. The evaluation stated that PP/IM should continue as a permanent international cooperative programme with several suggested amendments, both in contents, structure and further evaluation procedures, and a revised manual. The Executive Body decided in 1992 on the continuation of the programme under the name "International Cooperative Programme on Integrated Monitoring of Air Pollution Effects on Ecosystems" (ICP/IM) for the Convention.

Purpose of the programme and approaches to monitoring

Objectives of the programme
The main aim of integrated monitoring in the terrestrial environment is to determine and predict the state of ecosystems (or catchments) and their changes in a long-term perspective, with respect to the regional variation and impact of air pollutants, especially nitrogen, sulphur and ozone, and including effects on biota.

Ecologically speaking we might say that the aim is to differentiate between natural ecological variation plus succession, and anthropogenic perturbations caused by air pollutants in natural landscapes and ecosystems. The air pollutants embraced by the Convention on Long-range Transboundary Air Pollution should be specially emphasised. Hence, acidification through sulphur and nitrogen is given priority. Monitoring and evaluation of the effects of ozone, heavy metals, toxic organic substances and climatic change have also been suggested as parts of ICP/IM.

The objectives are:
1. To monitor the state of ecosystems and provide an explanation of changes in terms of causative environmental factors in order to provide a scientific basis for emission controls.
2. To develop and validate models for the simulation of ecosystem responses and use them; (a) in concert with survey data to make regional assessments, (b) to estimate responses to actual or predicted changes in pollutant stress.

The model approach is directed towards understanding long-term effects on biota. Present emphasis is on the determination of critical load and target load values of nitrogen and sulphur on terrestrial and surface water ecosystems. The use of undisturbed reference areas should make it possible to identify especially the net effects of long-range transported air pollutants on ecosystems and catchments.
3. To carry out biomonitoring for detecting natural changes, in particular to assess effects of air pollutants and climate change.

In a more long-term perspective the IM concept is useful among other things in monitoring loss of biodiversity, ecosystem effects of climate change and ozone depletion.

The ecosystem monitoring concept

Integrated monitoring of ecosystems means physical, chemical and biological measurements over time of different ecosystem compartments simultaneously at the same location. In practice, monitoring is divided into a number of compartmental subprogrammes which are linked by the use of same parameters (cross-media flux approach) and/or same/close stations (cause-effect approach).

Regional development of policy to regulate emission of anthropogenic pollutants (e.g. through development of critical loads) requires evaluation and assessment of environmental monitoring data. Assessment leading to policy definition is linked back to monitoring through the development and application of ecosystem models. The ICP/IM falls within the monitoring component of this overall framework, and the following discussion will focus on its specific position and role.

A national or international monitoring programme to evaluate the environmental effects of any anthropogenic perturbation (e.g. acidic deposition, toxic contaminants, climate change etc.) is best organised in an integrated, hierarchical manner. At the apex of the pyramid is a small number of intensively monitored process research sites. Here sufficient information is collected so that time-dependent models may be developed to predict future changes in the state of the ecosystem. The changes may occur in response to increased or decreased pollutant inputs. Many ECE countries operate a small number (1–10) of such sites.

Beneath the apex are regional monitoring networks, that use progressively less frequent sampling at progressively more sites. The base of the monitoring pyramid is composed of national "surveys" in which sampling may occur as infrequently as once or twice per decade. This number of hierarchical levels is probably a minimum for effective ecosystem monitoring on an international scale.

Within the hierarchy, the ICP/IM falls somewhat below the pyramid apex, and represents a source of information for comparison of complex and multiple effects across climatic gradients as well as geological, ecozone, and political boundaries. Much of the data reported to the international level are time averaged (e.g. monthly volume-weighted runoff concentrations). They are very useful for validating model and testing "universality". Once confidence in models performance has been obtained, application to lower hierarchical levels produces regional assessment, involving either temporal or scenario based production. Hence, multiple hierarchical levels of monitoring are necessary in order to supply the information needed for the model development-validation-application process. The IM presents the highest level having international cooperation and therefore, is in an excellent position to respond to the needs of international policy makers. On its own, however, the ICP/IM can not supply policy related information (e.g. critical loads); for political decisions we also depend on the simultaneous existence of lower hierarchies indicating the regional variation.

Two other features of the monitoring hierarchy should be noted. First, there should be some overlap between hierarchies to ensure data and model transferability among levels. Some ECE countries maintain one or more monitoring sites that contribute not only to process research but also to the ICP/IM and other ICP programmes. Such sites are the primary source of "ground truth" for validating and/or modifying ecosystem assessment models. Furthermore, it helps to maximise the scientific return obtained from the large resource expenditure required to operate such sites. Second, there is an inherent assumption of the continuing existence of all levels of the hierarchy. Piecemeal, intermittent, and short-term monitoring does not provide the information on temporal or spatial variations required to distinguish natural from anthropogenically induced effects. Arbitrary discontinuation of any given monitoring hierarchy may lead to collapse of the framework and an inability to effectively perform environmental assessment on either the national or international scales.

Mass balance performances
One of the central IM-approaches is to monitor the mass balance of major chemical components within the site. The approach consists of an open-system analysis of external fluxes. The aim is to quantify fluxes and to monitor the speed of changes in them. Simple mass balances can further be broken down into more complex ones for studying dose-response relationships.

Model applications
Prediction of the future response of ecosystems to changes in pollutant loading and environmental conditions is necessary from both a scientific and political viewpoint. These predictions provide the only basis for the formulation and quantification of remedial measures. In this respect, mathematical simulation models which are capable of predicting system response under future pollution deposition scenarios represent our best tools. These models must be capable of describing the physical, chemical and biological relationships observed in ecosystems The degree of damage to an ecosystem can then be estimated provided the models are based upon dose-response principles. Since the output from a model is only as good as the input data used to drive it, a comprehensive monitoring programme to identify the system function and provide adequate data for model calibration is essential.

Currently available models generally focus on one aspect of an ecosystem, notably atmospheric deposition, soil/soil solution chemistry or biology. Some of such models suitable to the ICP/IM for scenario testing are deposition models, hydrochemical models and biological models, although not all can be adequately parameterised at all sites in the ICP/IM. Initially, hydrochemical models will be utilised as the core of the modelling programme within ICP/IM as they have already been the subject of some quality control and testing. These models have also been validated to some extent and will provide reliable forecasts of the future changes in water quality which might be expected in relation to anthropogenic input of N and S. Biological models, on the other hand, are still in their infancy and require further development to achieve the mechanistic level of the hydrochemical and deposition models. Terrestrial models, incorporating plant growth, are currently under construction and will be capable of predicting long-term plant and vegetation response to changes in pollutant deposition. Aquatic models are currently based on empirical relationships between species diversity and survival and physical and chemical parameters of water quality. Nevertheless, these models, when linked to predictions from hydrochemical models, provide useful prognoses of future behaviour.

There is some way to go in model development before ozone and heavy metals are incorporated as driving variables into ecosystem models, and even the role of nitrogen is not fully understood. These developments must take place outside the ICP/IM. As new models are developed, however, they could be widely applied within the ICP/IM framework, as could all suitable existing models. The ICP/IM provides a unique database for validation and testing of such models, presuming complete data sets from the participating countries.

The comprehensive database from a few sites within the ICP/IM provides a unique opportunity for establishing links between models of individual ecosystem components. This will provide a powerful tool for the assessment of ecosystem response to future environmental change and the conceptual framework of the feedbacks and linkages of such a scheme are demonstrated in figure 1.

Sufficient data exists from many nominated ICP/IM sites to apply certain lumped models. The advantage of applying the same model to many sites is that a consistent approach can be utilised and sensible comparisons can be made. Once established, a model covering many sites can be used to evaluate emission control strategies, and long-term changes in policy, and used to investigate trends in the data.

The widespread coverage of sites in the ICP/IM is ideally designed for the application of models rather than model development. This is supported by the benefits o the central database allowing commonality of approach to data manipulation and aggregation for model calibration. Model development requires specific design of sampling and experimentation and the task is better left to more process oriented research programmes. The strength of the ICP/IM modelling effort lies in scenario assessment through widespread site application and the development of technologies for linking models for integrated assessment of environmental change utilising the integrated data sets available.

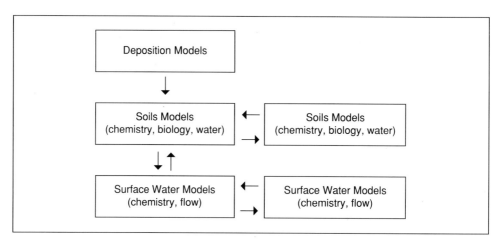

Fig. 1:
The Potential of Linking Models within the Framework of the Integrated Monitoring Programme.

The ICP/IM provides an essential database for model validation and prediction of ecosystem future response which is not available from other research programmes or sources, namely:

– The internally consistent and integrated hydrochemical and biological database will enable the future interaction between global climate change and atmospheric acidic deposition to be modelled and assessed.

– The integrated database will provide the platform for development of linked ecosystem models.

– The long time series of data will enable trend detection and model validation at a large number of sites and over a wide geographical area.

A number of opportunities exist at this stage of the ICP/IM to provide for an improved database from the point of view of modelling activities:

– Biological surveys will emphasised and must be undertaken on a regular basis if the aim of linking models is to be achieved.

– The basic data for applying models will be determined and must be measured at each intensive site and any gaps in the database will be addressed at the beginning of the permanent programme.

– The responsibility for model applications will be organised informally by several groups, not by a single modelling centre.

– Links with other international research and monitoring programmes will be formalised and maintained. As new models and processes are developed and identified this cooperation will ensure that any new relevant parameters are incorporated into the measurement programmes at IM sites and into the IM database.

Bioindication
Biological indications of environmental stress are important to recognise because they may serve as an early warning of ecosystem deterioration. Monitoring of biological variables makes it also possible to detect the cause-effect relationships within the ecosystem. One remarkable advantage of the ICP/IM is the possibility to integrate biological variables reliably to a wide selection of physico-chemical variables which are measured simultaneously. This is necessary if one tries to couple biological data in ecosystem modelling.

As the revaluation report of PP/IM (1992) states, forest growth and nutritional status are the most important variables from the modelling point of view. In addition to these, a collection of a number of self-indicating biological data are included which are not directly used in the models but can be used as indicators of changes.

There are also biological indices that may suit the framework of the ICP/IM but which are not found in the variable list of the programme. The reason is that the suitability of a variable for long-term monitoring depends also on advancement of methodology, cost of equipment and materials, availability of trained personnel and potential sources of funding. Still underdeveloped methods are one of the main problems when applying biological parameters to a monitoring system and for this reason many good indices cannot be used.

This information was extracted from the "Manual for Integrated Monitoring; Programme Phase 1993-1996" of the UN ECE Convention on Long-range Transboundary Air Polution, International Cooperative Programme on Integrated Monitoring on Air Pollution Effects. Additional information may be obtained from: Radovan Chrast, Air Pollution Section, Economic Commission for Europe, Palais des Nations, CH-1211 Geneva 10, Switzerland, Tel: +41 22 917 2358, Fax: +41 22 917 0123.

3.3 European Environment Agency (EEA)

The decision

The 29th October 1993 brought the long awaited decision on the location of the seat of the European Environment Agency (EEA). It will be established in Denmark; in the Copenhagen area. Preparations for the creation of the EEA are being undertaken by the EEA Task Force is DGXI.

Background and mandate

The creation of the EEA should be seen against the background of the aims of environmental protection laid down in the Treaty of Rome as amended bye the Single European Act by successive Community action programmes on the environment and now by the Maastricht Treaty. The United Nations Conference on Environment and Development (UNCED), and in particular Agenda 21, now provide the international framework for fostering cooperation on the environment.

The EEA and the 'European environment information and observation network', which it will coordinate, are intended to provide the Community and the Member States with objective, reliable and comparable information at the European level to enable them to take the measures necessary to protect the environment, as well as to be able to assess the results of these measures. Equally important is the aim to ensure that the public is properly informed about the state of the environment. For all these purposes the EEA will also provide the necessary scientific and technical support.

The Management Board will decide how the EEA is to be organised. However, it is clear from the Council Regulation that the EEA will be the hub of a decentralised distributed network making the maximum use of resources already existing in Member States.

Building blocks

A number of major building blocks are required to enable the tasks of the EEA to be carried out:

Information and Observation Network
The European environment information and observation network will consist of:
– component parts of the existing national information networks;
– a national focal point nominated by each Member State;
– institutions contracted by the EEA to work on specific subjects of particular interest, often referred to as "Topic Centres" or "Thematic Centres".

Within 6 months of the entry into force of the Council Regulation, Member States have to inform the EEA of the main components of their national networks, including, if desired, national focal points, as well as candidates for Topic Centres.

Multiannual Work Programme
The first multiannual work programme will be agreed by the end of July 1994.

Topic Centres
In approving the multiannual work programme the Management Board will have to decide specific tasks of Topic Centres.

Work areas

The principal activities will cover all those areas of work necessary to enable the EEA to describe the present and foreseeable state of the environment from the point of view of

- the quality of the environment
- the pressures on the environment
- the sensitivity of the environment.

Priorities
In supplying environmental information which may be used directly in the implementation of Community environment policy, the EEA in its first years will give priority to the following areas;

- air quality and atmospheric emissions;
- water quality, pollutants and water resources;
- the state of the soil, of the fauna and flora and of biotopes;
- land use and natural resources;
- waste management;
- noise emissions;
- chemical substances which are hazardous for the environment;
- coastal protection.

Special consideration will be given to transfrontier, plurinational and global phenomena, and the socio-economic dimension shall also be taken into account.

Possible future tasks
Within two years of the EEA coming into operation, the Council shall decide on further tasks for the EEA, in particular in the following areas:

- associating in the monitoring of the implementation of Community environmental legislation;
- preparing environmental labels and criteria for their award to environmentally friendly products, technologies and services, etc.;
- promoting environmentally friendly technologies and processes; and
- establishing criteria for assessing the impact on the environment.

Cooperation with bodies

Community bodies
The EEA shall actively seek the cooperation of other Community bodies and programmes, and notably the Joint Research Centre (JRC), the Statistical Office (EUROSTAT) and the Community's environmental research and development programmes.

The JRC will have an essential role in research and such areas as harmonization of environmental measurement methods; intercalibration of instruments; standardization of data formats; development of new environmental measurement methods and instruments; and other tasks as agreed between the Executive Director of the EEA and the Director General of the JRC. On request of the EEA, the JRC will be able to provide scientific support in specific fields of competence and technical support in areas such as informatics and networking.

Coordination with EUROSTAT is expected to lead to the supply of data on human activities (necessary to assess pressures on the environment), on economy-environment relations, and on basic territorial characteristics (topography).

Participation of non-Community countries and international organizations
In parallel with its cooperation with external bodies, participation of third countries in the work of the EEA is clearly important since environmental problems and challenges are not confined by national frontiers. Considerable interest has already been shown by EFTA (European Free Trade Association) member states, as well as by countries of eastern and central Europe.

Participation of international organizations
The EEA will also develop working relations with relevant international organisations, with bodies such as the Organisation for Economic Cooperation and Development (OECD), the Council of Europe and the European Space Agency, and those of the United Nations family, particularly the United Nations Environment Programme (UNEP), the World Meteorological Organization (WMO) and the International Atomic Energy Agency (IAEA). The participation of non-governmental organisations is equally important.

The inheritance

The EEA is the logical successor in particular of the CORINE programme (COoRdination of INformation on the Environment). CORINE was an experimental project (1985-1990) to determine the need and practice for collecting, coordinating and ensuring the consistency of information on the state of the environment and natural resources within the Community. Its pilot projects concentrated on land cover, biotopes and emissions to the atmosphere (see below).

The CORINE programme led to three types of complementary results: (i) relevant data bases for several priority environmental concerns; (ii) methods and nomenclatures adopted at Community level and internationally; and (iii) networks of experts from all Member States used to pooling their knowledge for environmental objectives. The techniques, data bases and networks established under the CORINE programme will provide a significant contribution to the work of the EEA.

The conclusions of the Dublin, Dobris and Lucerne Ministerial Conferences (June 1990, June 1991 and April 1993 respectively) emphasised the importance of integrating environmental information systems throughout Europe. Many non-EC countries have already expressed a keen interest to take part fully in the EEA's programme and to integrate into the network. In particular, the EFTA countries have stressed the importance of the role of the EEA as an instrument of cooperation on a European scale.

Beginning in January 1990 the EEA Task Force has worked to maintain and utilise the data expertise built-up during the CORINE programme. and to handle the many and diverse initiatives arising out of the decision to establish the EEA. The main areas of the Task Force's activities in the period 1990 to 1993 are outlined below.

Preparatory activities for the EEA

Work programme
In the light of many discussions between the Commission and a high-level group of Member State national experts, the EEA Task Force has prepared a first draft multiannual work programme. A systematic analysis of the data needed to implement the programme has been started, based in the first place on the objectives of the 5th Community Action Programme on the Environment, 'Towards Sustainability' (Council Resolution, 1.2.1993) and taking into consideration the reporting requirements established in the UNCED Agenda 21.

Network
Work is underway to develop a network architecture for the EEA to ensure inter-operability of the various elements of the 'European environment information and observation network'. This is taking into account issues of security, secrecy and legality, in line with common rules developed through the DG III programme, IDA (Interchange of Data between Administrations), whose aim is to encourage and stimulate such data exchange. The approach focuses on the requirements for communication and the exchange and access to data and information systems.

Databases
A 'data-model' is being prepared to serve the needs of current and expected future data holdings including, in particular, those arising out of European environmental legislation. Development of a strategy for coding data, and standards needed to improve and facilitate data exchange, are included.

Catalogue of data sources
A report has been completed on the needs and possible solutions for a 'meta-database' for the EEA. This is a system that records details of data holdings, dada sources, institutions. activities and other environmental related information needed for the EEA's work. The main characteristics of the system have now been defined and a standard 'data-model' is being developed. Plans are also being laid for the preparation of a multi-lingual thesaurus of environmental terms to use in connection with the catalogue.

Updating and completing CORINE data bases

Land cover
The cartography of Land Cover on the scale of 1:100 000 has been completed for almost half of the territory of the European Community. This is continuing with the support of Community regional policy, with a view to completion of the inventory in 1994-95. In addition, the application of Community methodology has begun in the PHARE countries (Countries of central and eastern Europe in receipt of EC economic aid according to the terms of Council Regulation 89/3906/ECC), as well as in several EFTA member states and in countries in the southern part of the Mediterranean basin with LIFE/MEDSPA support.

Biotopes
The data base has been updated and supplemented to serve the needs of Community nature

conservation policy, and in particular the application of the 'Habitats' Directive (Council Directive 92/43/ECC of 21 May 1992 on the conservation of natural habitats of wild fauna and flora). The PHARE countries and certain EFTA member states are carrying out the inventory according to the Community methodology. Work continues in collaboration with the Council of Europe and the principal international nature conservancy organisations.

Corinair 1990
An inventory of pollutant emissions is being carried out for 1990 as an update of the 1985 inventory (COM (88) 420 final of 22 July 1988 and Corinair report EUR 13232 FR). The work is being performed in cooperation with the United Nations' Economic Commission for Europe (UNECE), all of whose European member states currently apply the Community methodology 'Corinair'. The inventory includes greenhouse gases, and the results will provide the European contribution to the OECD-IPCC: (Intergovernmental Panel on Climate Change) programme in this field. The inventory also meets the needs for the supply of data connected with the Council Directive on large combustion plants (Council Directive 88/609/EEC of 24 November 1988 on the limitation of emissions of certain pollutants into the air from large combustion plants). A feasibility study has been completed which confirms that the Corinair methodology should be extendible to other environmental media to build an integrated multi-media inventory of emissions.

Preparation of the report Europe's Environment 1993

The Task Force is completing preparations of the pan-European state of the environment report (Europe's Environment. The Dobris Assessment). The report is being prepared by the EEA Task Force in collaboration with the UNECE and other international bodies (Council of Europe, IUCN, OECD, UNEP, WHO). It will provide an assessment of the state of the whole European environment and of the pressures caused by human activities and an analysis of prominent environmental problems of concern for Europe. The report will be accompanied by a statistical Compendium (being prepared with EUROSTAT, OECD, UNECE and WHO). Other products such as an environmental atlas and a popular version will appear later.

This information is extracted from a paper submitted to UNEP-HEM for the workshop. Additional information may be obtained from the Commission of the European Communities, DG XI Environment, Nuclear Safety and Civil Protection, European Environment Agency Task Force, rue de la Loi 200, B-1049 Brussels, Tel: +32 2 29 68811, Fax: +32 2 29 69562.

3.4 Arctic Monitoring and Assessment Programme (AMAP)

Introduction

At the Ministerial conference in Finland in 1991, the Ministers from the eight Arctic countries agreed to develop an Arctic Monitoring and Assessment Programme (AMAP). In the strategy document from the Ministerial Meeting it is stated that *"The primary objectives of the programme are the measurement of the levels of anthropogenic pollutants and assessment of their effects in relevant compartment parts of the Arctic environment. The Assessment should be presented in status reports to relevant fora as a basis for necessary steps to be taken to reduce the pollution"*.

The objectives for AMAP are as follows:

– to monitor, assess and report the status of arctic environment

– to document and assess the effects of anthropogenic pollution

– to recognize the importance, relationship to, and the use of the arctic flora and fauna by the indigenous peoples

– to document levels and trends of pollutants

– to document major sources and processes important in the transportation, precipitation and accumulation of pollutants

To implement the AMAP an Arctic Monitoring and Assessment Task Force (AMAP TF) (at the ministerial meeting in Nuuk, 1993 the Task Force was substituted by a Working Group, AMAP WG) and a small permanent secretariat were established by the government of Norway. The First AMAP WG meeting was held at Tromso, Norway in December 1991.

AMAP WG consists of representatives from each of the Arctic Countries which signed the Declaration on the Protection of the Arctic Environment. It will meet at least once a year and may set up workshops where necessary for the performance of tasks which will be determined. Representatives from the following organizations representing indigenous people have been nominated as observers to the AMAP WG: The Inuit Circumpolar Conference, The Nordic Saami Council and The USSR Association of Small Peoples of the North. AMAP WG may also invite observers representing Non-Arctic States and international organizations involved in significant monitoring and research that is of relevance to the Arctic. At the first Working Group Meeting the following organizations and countries participated as observers: IASC, UNEP, UN ECE, ICES, United Kingdom, Poland and the Federal Republic of Germany. Figure 1 gives an overview of the observers to the AMAP WG.

The Secretariat will coordinate the Task Force activities between sessions, act as a secretariat at the Working Group meetings and where necessary, at meetings of its subsidiary working groups. The Secretariat shall establish a reporting system for discharges and emissions, including spills. The Secretariat will play a major role in coordinating the preparation of the State of the Arctic Environmental Reports and in coordinating the exchange of scientific data between participating states.

Indigenous Organizations:
- Inuit Circumpolar Conference
- Nordic Saami Council
- The Russian Association of the People of the North

International Organizations:
- International Atomic Energy Agency (IAEA)
- International Arctic Science Committee (IASC)
- International Council for the Exploration of the Seas (ICES)
- Nuclear Energy Agency (OECD/NEA)
- Oslo and Paris Commission (OSPARCOM)
- United Nations Environment Programme (UNEP)
- Economic Commission of Europe (UN ECE)
- World Metrological Organization (WMO)
- Norther Forum
- International Union for Circumpolar Health (IUCH)
- The Arctic Environmental Protection Strategy (AEPS) working group; CAFF, PAME and EPPR

Countries:
- Germany
- Poland
- The Netherlands
- United Kingdom

Fig. 1:
Observing Organizations and Countries

AMAP – The monitoring programme

Norway was the lead country in preparing a draft proposal for the AMAP. The first draft was presented in 1990, and an updated draft discussed at an Expert Meeting arranged in Oslo in November 1990 where more than 70 experts from the participating countries attended. Based on the recommendations from this Expert Meeting, a proposal was presented to the Ministerial Meeting in 1991. During 1992 the monitoring programme was upgraded by the use of lead countries: Canada had the responsibility for the atmoshperic sub-programme, Denmark the human health, Finlanf the freshwater, Norway the marine and Sweden the terrestrial. The final programme for the first period of AMAP was decided upon in 1993, published as an AMAP report 93:3. The programme specifies the:

- parameter to be monitored in each compartment of the actual media, e.g. in sediments, in water, in animals and which of the biological effects parameters that are to be monitored;

- methodology to be followed for sampling, storing, pre-treatment/extractions, analysis and reporting, including proposals for the number of replicates to be sampled, analysed etc.;

- quality assurance programmes to be followed while performing the sampling, storing, pre-treatment/extraction, analysis and reporting;

- ideal geographical arrangement of the sampling stations so that it will be possible to present an assessment that is reliable. The stations should be arranged in a way that will enable the programme to cover known polluted areas, areas expected to be "sedimentation or precipitation areas" and areas expected to be "non-polluted" or background areas;

– sampling frequency, because for some parameters the natural variation may be unknown and it is important to have this variation documented before an assessment can be performed and the trend monitoring initiated.

The Ministers have also requested AMAP to establish a regular system for the reporting of emissions and discharges. AMAP intend to base this work on existing national and international reporting systems, e.g. the Oslo and Paris Commission and UN-ECE EMEP.

The AMAP programme has as far as possible been built on already existing programmes that are either running within the Arctic, or in adjacent areas. This means that the same parameters, methodology, quality assurance recommended for existing programmes e.g. EMEP, North Sea Task Force, the UN ECE Integrated Monitoring Programme, etc, have been followed by AMAP. By building on existing programmes, it should be possible to compare the results from the different programmes and present overviews that cover greater areas from the mid latitudes and up to the Arctic. However, necessary adjustments in the methodology may need to be performed so that the methods are applicable to the arctic conditions.

AMAP has tried to harmonize existing national programmes, but has also initiated new programmes to fill in gaps in knowledge where necessary.

AMAP has been initiated through a step by step procedure. As an initial priority, AMAP has focused on persistent organic contaminants, selected heavy metals, radionuclides, and ultimately the monitoring of ecological indicators to provide the basis for the assessment of the status of the arctic ecosystems.

The first programme for AMAP also contains recommendations for the monitoring of acidification, eutrofication, hydrocarbons other than PAH, and radiatively important trace species (RITS), e.g. greenhouse gases. Two of the most significant threats to the present Arctic environment may come from climate change and the depletion of stratospheric ozone. Monitoring programmes have already been established to detect and determine the causes and effects of climate changes and ozone depletion. It is therefore important for AMAP to be aware of these programmes (e.g. the World Climate Research Programme, the International Arctic Buoy Programme, etc.) and to develop links with them from an Arctic perspective in order to encourage and facilitate an Arctic component in these programmes. Data obtained for assessing climate change will provide important inputs to the AMAP dataset. The study of biological effects is an important part of AMAP and effects such as vegetation defoliation, and species composition/abundance etc. may be due to both direct pollution and climate changes. Thus the AMAP data and assessments will be relevant to climate change programs in the Arctic.

At the Task Force Meeting in Tromsø, it was decided that the monitoring programme should be performed along media specific lines and not pollution specific lines. Even so, it is the intention that the monitoring programme shall be able to follow a pollutant from the discharge, through the transport process to the sedimentation/ precipitation, or through it's accumulation in biota ending up in the highest trophic level, the human.

AMAP has constructed a Project Directory (AMAP PD) that contains information about research and monitoring activities in the Arctic area. At present 530 projects have been reported, among which 312 are nominated as national implementation plans for eight arctic countries and the observing countries. For these 312 projects there exists detailed descriptions of parameters, methods, stations, etc. The AMAP PD will be updated twice a year and distributed on request. Part of it will also be available on-line through GRID-Arendal who have been involved in this project. The AMAP PD will be a major contributer to the new Arctic Data Directory (ADD) that is under construction between USA, Canada, Nordic countries and Russia.

At the First Task Force Meeting it was decided to delete the special group on climate. Due to this the monitoring of snow and ice have to be covered by the other groups, e.g. sea ice by the marine groups etc. Remote sensing (using satellites, airplanes, buoys etc.) may play an essential role in the monitoring of great areas of the Arctic. USA was appointed to be the lead country to evaluate where remote sensing can be used in the monitoring. In addition the USA is the lead country regarding the use of modelling in AMAP, e.g. as a part of the planning and assessment of the data. Within some areas, e.g. atmospheric and marine environment, models are commonly used. It is the intention to establish some models that could cover greater areas of the Arctic and be used both for describing the processes involved in the transportation of pollutants into the Arctic area and to potentially foresee the effects of any reductions in discharges or emissions.

AMAP – The assessment

AMAP has developed a specific strategy for how to perform the assessment. First the monitoring programme was designed to answer specific questions of interest for the assessment. For the assessment work that starts during 1994 Lead Countries have been appointed to take the responsibility for the completion of a given chapter, Canada and Sweden for persistent organics, USA and Russia for heavy metals, Norway and Russia for radioactivity, Finland for acidification and Denmark for human health. A small group of experts is responsible, but they will be assited by national experts in drafting of specific parts of the chapters. The secretariat together with the Assessment Steering Group (ASG) will be responsible for the drafting of the introductory and concluding chapters. The drafts will be circulated for national comment during 1995–96. All the leaders in these drafting groups, the AMAP Board, one representative from the indigenous organizations and Iceland are members of the ASG that shall oversee the process and together with the AMAP secretariat coordinate all work related to the assessment.

AMAP will prepare two reports. One comprehensive technical and scientific assessment report on the status of the Arctic environment, and a second report that will be shorter and is intended for the Ministers and the general public.

The assessment will be based on information already printed in international journals, new data created through AMAP´s monitoring programme and national reports. For the compilation and quality control of new data AMAP has made contracts with international thematic data centres to handle data from the implementation plans, e.g. with the International Council for the Exploration of the Seas (ICES) in Copenhagen that will handle all marine data, and The Norwegian Institute for Air Research (NILU) that will handle the atmospheric data. A new database for handling the radioactivity data from the atmosphere, terrestrial and aquatic areas is under development as a response to a specific request from the ministers. For the handling of terrestrial and freshwater data AMAP is still seeking possible institutions. The final report shall be presented to the ministers in 1997 and this will include recommendations for actions.

The observers

Representatives from the indigenous organization for people living in the Northern areas have been involved in the process to establish AMAP and they will have an important role in the assessment by providing data and taking part in the assessment. In addition some international organizations and countries that are involved in arctic research have been invited as observers to the Working Group (fig. 1).

The Arctic is a huge region and to cover this area with a satisfactory number of sampling stations is an enormous task. It is therefore hoped that Non-Arctic countries performing research in the Arctic can also contribute to the monitoring, e.g. by monitoring AMAP parameters at their fixed stations or on their research cruises.

AMAP – The time schedule

The field monitoring was performed during 1992–1994, and will to some extent continue in 1995 at least for the priority compounds and parameters considered to be essential. For some parameters some monitoring stations have already been established within the Arctic, and monitoring has been going on for several years e.g. for atmospheric pollutants and climate changes. Monitoring will be initiated in some other areas, e.g. the joint monitoring of the Barents Sea and Kara Sea between Norway and Russia. The methodology for these monitoring programmes will be to a great extent in accordance with guidelines and manuals endorsed by other international programmes and be accepted by the AMAP WG as the methodology to be used by the AMAP.

Closing remarks

The proposal of a joint monitoring programme for the Arctic has met with a very positive response from all countries involved. It is remarkable that the international society has succeeded to establish such a programme over such a short time. All participating countries have expressed their positive willingness to do the necessary tasks and it is hoped that this will be reflected in the national budgets for AMAP. The monitoring will cost a lot of money and as a principle each country shall finance its own activities. It is the hope that this positive attitude will continue so that AMAP can be performed on a sound scientific basis and that the documentation of the results are so good that in the end long discussions and questions regarding the validity of the data presented in the status report and the representativeness of the samples ca be avoided.

This information is extracted from a paper submitted to UNEP-HEM for the workshop. Additional information may be obtained from Lars-Otto Reiersen, AMAP, P.O. Box 8100 Dep, N-0032 Oslo, Norway. Phone: +47 22 573544, Fax: +47 22 676706, e-mail: l.o.reiersen@amap.uio.no

3.5 Environmental Monitoring and Assessment Programme (EMAP), USA

The Environmental Monitoring and Assessment Program (EMAP) is an interdisciplinary, multi-agency programme designed and initiated through the U.S. Environmental Protection Agency's (EPA) Office of Research and Development. It is an ecological research, monitoring and assessment programme with national and regional scope, which is integrated and scientifically based and designed to address important questions about our environment. The program's objectives and scale require that EMAP be an interagency program, with ongoing, active participation and involvement by many Federal agencies, such as the U.S. Department of Agriculture's Agricultural Research Service, Soil Conservation Service, and Forest Service; the U.S. Department of the Interior's Fish and Wildlife Service, Bureau of Land Management, National Park Service, Geological Survey and National Biological Survey; the Department of Commerce's National Oceanic and Atmospheric Administration; and others.

EMAP demonstrates EPA's ongoing efforts to change the way it does business: to inject science more prominently into decision-making and to help focus the Agency's resources on those problems that pose the greatest risk to the environment. It has already begun to make scientific contributions to management decisions. These contributions will continue to accrue with time. EMAP addresses the large scale, longer-term environmental problems occurring at regional and national scales and, thereby, complements the local scale, shorter term monitoring programmes within State and local agencies. EMAP should be viewed as an integral part of our environmental protection and management activities into 21st century, not as a short-term solution to current problems.

EMAP: Monitoring for results

EMAP evolved from discussions about basic elements needed in a monitoring program to contribute to decision-making on environmental protection and resource management. These elements included

1. A focus on social values and policy-relevant questions.

2. Approaches that assess and translate scientific results into information useful for decision-makers and the public.

3. Ecological indicators for monitoring the condition of key ecological resources rather than individual pollutants or stressors.

4. Periodic estimates, with known confidence, of the status and trends in indicators of ecological condition.

5. An integrated approach to monitoring that includes major ecological resources.
6. Implementation with regional scales of resolution, rather than an individual site or local area orientation.

7. An interagency, interdisciplinary program in which all participating agencies are cooperative partners in the research, monitoring, and assessment efforts.

Goals and objectives

EMAP's goal is to monitor and assess the condition of the Nation's ecological resources, thereby contributing to decisions on environmental protection and resource management. To accomplish this goal, EMAP works to attain four objectives:

1. Estimate the current status, trends, and changes in selected indicators of the Nation's ecological resources on a regional basis with known confidence.

2. Estimate the geographic coverage and extent of the Nation's ecological resources with known confidence.

3. Seek associations between selected indicators of natural and anthrophogenic stresses and indicators of condition of ecological resources.

4. Provide annual statistical summaries and periodic assessments of the Nation's ecological resources.

Values and questions

There are three general perspectives on values that relate to ecological resources:
1. Social values which incorporate the broadest spectrum of environmental goals and values desired for ecological resources and expressed through the legislative process;

2. Administrative values which include the management – regulatory agencies and their legislative mandates to protect and manage both specific ecological resources and the total environment; and

3. Scientific values which incorporate scientific questions, principles, and knowledge of ecological structure and function with an understanding of ecological responses to human disturbances.

Environmental and resource management decisions require that available information address values and questions based on these perspectives. EMAP's results should provide useful information to legislative, administrative, scientific, and public users. To serve this diverse group, EMAP must continually focus on environmental values and questions important to them. EMAP will not establish environmental policy, regulatory, or management strategies, but it must provide information that contributes to forming and evaluating theses strategies. This interactive process requires continuous feedback to users to provide scientific information and procedures useful in answering their questions.

Identifying values and the associated questions relevant to these perspectives is an important first step in the EMAP process because it provides a direct link to the user. Values desired for ecological resources typically fall into three categories:

1. Biological integrity – the ability to support and maintain a balanced, integrated, adaptive community with a biological diversity, composition, and functional organization comparable to those of adjacent "natural" and manged systems in the region.

2. Consumptive use of value - worth of environmental goods or services based on their ability to be *extracted*. "Worth as used here is not necessarily intended to denote monetary value of a resource. Consumptive use of values involve extracting something from natural systems in order to produce a tangible product of worth to humany. Examples include fishing, hunting, timber harvesting and agricultural production. [Note: *extractive values* also include values generated form the disposal of waste materials into the environment as a by-product of the production of conventional goods and services] (Bishop and Hoag, 1993; Freeman, 1993; Scodari, 1990, as modified).

3. Non-consumptive use value – the worth of environmental goods or services, which are not extracted. or are only passively exploited, by those holding value for this good or service. "Worth" as used here is not necessarily intended to denote monetary value of a resource. Non-consumptove use values involve all values not derived via extractive activities. Examples include hiking and birdwatching, as well as "non-use" or "passive use" values associated with the mere existence of natural assets, called *existence values* [i.e., values based on bequests to future generations, and motives other than personal enjoyment of resources of products derived from resources (e.g., altruism, stewardship, moral convictions)] and aesthetic or *amenity* values (e.g., scenic value, spiritual value) [Note: *option value* (having the option to use the good or service later) should not be applied in a list of non-use values because, to the extent that it is relevant, it is already considered] (Bishop and Hoag; Freeman, 1993, Scodari, 1990, as modified).

EMAP is designed to address questions that relate to attributes of a population (in a statistical sense) of an ecological resource. The EMAP focus is not on individual lakes or streams or forest stands in a region but rather on characteristics of interest for the total number of lakes, miles of streams, and acres of forest in a region. This focus is compatible with the scale at which Federal programs typically operate.

EMAP's integrated approach

EMAP will compare the status and trends in ecological condition among multiple ecological resources and assess the cumulative effects of environmental stresses on these resources. EMAP will assess these effects by integrating measurements within and across different classes of ecological resources, for example, bottomland hardwood, small estuaries, rivers, deserts. Integration refers to 1) combining, linking, and analyzing data from all relevant ecological resources, media, and monitoring networks; 2) ensuring the quality of these data at an acceptable level; and 3) using these data in ecological assessments to develop a holistic perspective of the condition of the Nation's ecological resources and possible factors contributing to this condition. In addition, EMAP has developed an integrated strategy for its monitoring and coordinating components.
All resource groups within EMAP are using compatible sampling designs, conducting annual field surveys, and interacting with other agencies that conduct monitoring programs. Further, EMAP has coordinated activities to ensure that resource groups follow consistent, compatible, and comparable strategies for indicator development, information management, quality assurance, methods development, logistics, and assessment and reporting. Finally, the concept of adding value or assessing the information in a policy-relevant context represents a central theme underlying all EMAP activities.

Resource monitoring

EMAP will monitor major categories of ecological resources: 1) agricultural lands, 2) rangelands, 3) estuaries, 4) forests, 5) the Great Lakes, 6) surface waters, lakes asnd streams. Wetlands will be investigated under the relevant ecolsogical reosurce. EMAP will also conduct research on the interaction of these resources on the landscape (Landscape Ecology). EMAP's seven resource groups are named after these resource categories (including Landscape Ecology).

For these resource categories, EMAP will monitor selected indicators of ecological condition and will collect and compile data on selected stressor indicators, including climate and atmospheric deposition. The program will integrate its monitoring of indicators within and across resources, such as forest, surface waters, and wetlands, so that researchers can detect changes in indicators of ecological condition at large spatial scales over time. Large-scale integration represents one of the greatest technical challenges in EMAP.

Sampling design

The statistical approach being implemented in EMAP is similar in concept to other Federal statistical programs or surveys. A principal difference is that these programs focus on producing estimates of characteristics for human populations, business establishments, or agricultural enterprises rather than ecological resource populations. In contrast, EMAP focuses on producing estimates of attributes from ecological resource populations such as prairie pot-hole wetlands, the Great Lakes, grasslands in the Great Basin, or forest lands in the United States.

To address EMAP's objectives, regional populations of all major ecological resources in the United States are emphasized, not individual ecosystems. The design permits estimates of the condition, geographic coverage, and extent for regional populations of ecological resources. The design permits population estimates to be provided with known confidence, i.e., statistically defensible, quantitative statements of uncertainty must accompany the estimates. EMAP requires these estimates not only for a specific point in time (current status) but also repeated over time (trends). The design enables associations (empirical relationships) to be investigated between condition indicators and stressor indicators for the ecological resource.

To achieve its objectives, EMAP uses a probability-based sampling design over time and space as a cost-effective monitoring program (Overton et al. 1990). EMAP'S sampling design uses a specific pattern for repeated sampling of sites over time and a systematic grid structure as a basis for distributing the sample sites over space. These two basic procedures are implemented jointly, not independently, to further enhance the cost-effectiveness of the selected approach (Overton et al. 1990). A systematic grid superimposed over the entire United States is the basic structure used to implement the sampling design over time and space.

Ecological indicators

To assess status, changes, and trends in the condition and extent of the Nation's ecological resources, EMAP will monitor ecological indicators (Bromberg 1990, Hunsaker and Carpenter 1990, Hunsaker et al. 1990). Indicators are defined as any characteristic of the environment that can provide quantitative information on the condition of ecological resources, magnitude of stress, exposure of a biological component to stress, or the amount of change in condition.

EMAP's strategy will be to emphasize indicators of ecological structure, composition, and function that represent the condition of ecological resources relative to social values. Through rigorous scientific research, EMAP is selecting, developing and evaluating indicators that describe the overall condition of ecological resources; permit the detection of changes and trends in this condition; and provide preliminary diagnosis of possible factors that might contribute to the observed condition, such as human-induced versus natural stressors. The program emphasizes the development and evaluation of biological indicators.

EMAP defines two general categories of ecological indicators: condition and stressor indicators. A condition indicator is any characteristic of the environment that provides quantitative estimates on the state of ecological resources and is conceptually tied to a value. There are two types of indictors: biotic and abiotic. EMAP will estimate the regional distribution of quantitative values for each of these indicators within and among ecological resource categories. All estimates will be accompanied with specified levels of confidence so the user knows the certainty of the estimates.

Stressor indicators are characteristics of the environment that are suspected to elicit a change in the condition of an ecological resource, and they include both natural and human-induced stressors. Selected stressor indicators will be monitored in EMAP only when a relationship between specific ecological condition and stressor indicators are known, or if a testable hypothesis can be formulated. Associations between selected indicators of stress and ecological condition can provide insight and lead to the formulation of hypotheses regarding factors that might be contributing to the observed condition. These associations can provide direction for other regulatory, management, or research programs seeking to establish causal relationships.

Assessment

Assessment is the process of interpreting and evaluating EMAP results for the purpose of answering policy-relevant questions about ecological resources. It includes determining the fraction of the population that meets a socially defined value or relating associations among indicators of condition and stressors.

Assessment includes several different stages of analysis along a continuum of increasing complexity. EMAP contributes directly to the first four stages in the continuum and indirectly to the latter three.

1. Current status – The first stage involves analyzing selected indicators to describe the status in ecological condition of resources at a particular time or during a particular period.

2. Detection of change – The next stage is the examination of changes and trends in selected indicators of condition and extent.

3. Evaluation of the significance of change in condition – Going beyond the statistical characterization of resource condition are issues relating to the significance of change with regard to values.

4. Association of change/stress – The fourth stage of analysis is establishing statistical associations between spatial/temporal patterns in selected indicators of stressors and condition.

5. Establishment of causality – Establishing cause-and-effect relationships between specific changes in ecological indicators and particular anthropogenic stresses is a fifth stage. These analyses include assessing interactions among multiple anthropogenic stresses and natural variability. EMAP is not a program to determine cause and effect, but it should be able to associate the ecological conditions with possible stressors to guide and direct other research in determining the causes of these responses.

6. Predictive capability - Each of the previous stages of analysis retrospectively utilizes historical and current monitoring data to establish change and association. Predictive capability is intrinsically prospective and requires the development of predictive tools that go beyond monitoring and retrospective assessment. EMAP will estimate past and present trends in resource condition and provide information against which to compare predictions of ecological conditions.

7. Ecological risk assessment – This is a much broader set of activities that includes problem formulation and ecological effects, exposure and risk characterization (RAF 1992). Ecological risk assessment represents one of the fundamental ways EPA is attempting to change the way it does business. EMAP contributes to formulating problems and translating scientific information to address these problems within EPA's framework for ecological risk assessment.

Implementation

Because of scientific, logistical, and funding constraints, EMAP is being implemented in phases that occur both within and among ecological resources and among geographic regions. As a result, implementation generally progresses through four phases: a pilot project, then a demonstration project, then regional implementation, and finally national implementation.

EMAP will conduct a pilot multiple resource study in the Mid-Atlantic region of the Eastern U.S. This geographic initiative Mid-Atlantic Integrated Assessment (MAIA) project includes the land area and near coastal waters associated with all of the states of Pennsylvania, West Virginia, Maryland, Delaware, Virginia and adjoining drainage basins in the states of New Jersey, New York, and North Carolina. Primary components of the regional study will be : (1) a monitoring dimension that provides data on all ecological resources at some geographic scale; (2) a research dimension that establishes the scientific foundation for understanding and predicting; (3) an assessment dimension that translates scientidic results into information useful for decision-making.

The project will demonstrated the full power and utility of the EMAP programme and design, at least with respect to the condition and status of the condition of ecological resources. The assessment will employ the EMAP design and data collection activities, while implementing conjunctive research efforts. Research questions to be addressed relate to issues such as spatial scaling, EMAP design, landscape/ land use patterns and resource conditions, validity of a EMAP conceptual framework, common values across resources and common metrics for ecological indicators, use of ecolcogical condition indexes, and validity and utility of ecological risk assessment methods. Ecological assessment and other methods will be field-tested and utilized to examine cause-effect relationships that explain resource condition and to characterize risks.

The purpose of a multiple resource ecological assessment, then, will be to better understand how interrelationships among categories of ecological resources are important in characterizing overall ecosystem health and how aassessment science can contribute to explaining causal linkages. Questions about ecosystem health will be examined using ancillary stressor data and ecological risk assessment methods. Results will provide insight into the efficacy of "ecosystem" and "watershed" management approaches to addressing resource and environmental protection concerns.

References

BISHOP R. C. and Hoag D. (1993). EMAP and Policy Analysis: Giving Assessment Primacy. Unpublished white paper commissioned by the U.S. Environmental Protection Agency, Office of Research and Development, to the Resource Policy Consortium, Contract No. 2D3238NAEX
BROMBERG S. M. (1990). Identifying ecological indicators: An environmental monitoring and assessment program. Journal of the Air Polllution Control Association 40: 976–978.
FREEMANN A. M. III (1993), The Measurement of Environmental and Resource Values: Theory and Methods. Washington, DC: Resources for the Future.
HUNSAKER C. T. and CARPENTER D. E. (1990). Environmental Monitoring and Assessment Program: Ecological Indicators. EPA 600390060. Washington, DC: U.S. Environmental Protection Agency.
HUNSAKER C. T., CARPENTER D. E. and MESSER J. J. (1990). Ecological indicators for regional monitoring. Ecological Scociety of America Bulletin 71:165–172.
OVERTON W. S., WHITE D. and STEVENS D. L., Jr. (1990). Design Report for EMAP. EPA 600391053 Corvallins OR: U.S. Environmental Protection Agency, Environmental Protection Laboratory.
RAF (Risk Assessment Forus) (1992). Framework for Ecological Risk Assessemnt. EPA 630R92001. Washington, DC: U.S. Environmental Protection Agency.
SCODARI P. (1990). Wetlands Protection: The role of Economics. Washington, DC: Environmental Law Institute.

This information is extracted from the Environmental Monitoring and Asessment Program Guide, October 1993. Additional information may be obtained from: EMAP Research and Assessment Center, Environmental Monitoring and Assessment Program, Office of Research and Development, U.S. Environmental Protection Agency, Research Triangle Park, NC 27711.

3.6 Long-Term Ecosystem Research (LTER), USA

With an initial set of six sites selected in 1980, the National Science Foundation established the Long-Term Ecological Research (LTER) Program to conduct research on long-term ecological phenomena in the United States. The present total of 18 sites represents a broad array of ecosystems and research emphases. The LTER Network is a collaborative effort among over 600 LTER scientists and students which extends the opportunities and capabilities of the individual sites to promote synthesis and comparative research across sites and across ecosystems.

The mission of the LTER Network is to conduct and facilitate ecological research by

- Understanding general ecological phenomena which occur over long-term and broad spatial scales
- Creating a legacy of well-designed and documented long-term experiments and observations for use by future generations
- Conducting major synthesis and theoretical efforts
- Providing information for the identification and solution of societal problems

Intersite research and synthesis

In the commonalities of its sites, the LTER Network creates greater scientific, social and administrative opportunities to take a broader view than most other ecological research programs. Current synthesis efforts include work in process studies, climate forcing, analysis of temporal and spatial data, and scaling up to continental and global scale.

LTER scientists

- Ask similar scientific questions in a wide variety of landscape
- Have similar sets of measurements in core research areas, and accessible data which can be shared
- Have access to and support for incorporating new technologies
- Have mechanisms for communication, planning and data sharing
- Regularly associate with other ecologists, broadening their exposure to different research approaches and ideas

Research opportunities at LTER sites

The LTER Network offers the broader environmental biology research community, including students and foreign scientists, the opportunity to use the sites for both long- and short-term projects appropriate to individual sites, a group of sites, or the Network as a whole. Initial arrangements for collaborations should be made with personnel at the site(s) under consideration, and proposals to the National Science Foundation for such collaborative work should be submitted to the relevant disciplinary program. Program directors may also be contacted for information about short-term funding opportunities.

LTER sites share a common commitment to long-term research on the following core research topics:

- Pattern and control of primary production
- Spatial and temporal distribution of populations selected to represent trophic structure
- Patterns of inorganic inputs and movements of nutrients through soils, groundwater and surface waters
- Patterns and frequency of site disturbances

National and international activities

As a global leader in long-term research and monitoring activities, the U.S. LTER Network has established linkages with existing and developing domestic long-term ecological research programs, as well as similar programs around the world. These relationships range from exchanges at the individual scientist and site research program levels to participation in international meetings to global-scale research planning and collaboration. In the next decade, the LTER Network plans to undertake a major expansion of its national and international activities.

National

- Developing strong multidisciplinary science and public education programs

- Developing a multi-agency effort to add sites for wider representation of key biomes and portions of major gradients

- Creating synthesis centers at selected LTER sites

- Facilitating the advancement of current ecological science and innovative research

International

- Assisting in the establishment of networks for long-term ecological research in other countries

- Creating programs between U.S. and foreign LTER sites and networks

- Developing and operating a communications and data sharing system among an international network of sites

- Facilitating the establishment of a global system of environmental research sites

LTER major ecosystems

1. H.J. Andrews Experimental Forest, Oregon
   Temperate coniferous forest
   Research topics: Successional changes in ecosystems; forest-stream interactions; population dynamics of forest stands; patterns and rates of decomposition; disturbance regimes in forest landscapes.

2. Arctic Tundra, Alaska
   Arctic tundra, lakes, streams
   Research topics: Movement of nutrients from land to stream to lake; changes due to anthropogenic influences; controls of ecological processes by nutrients and by predation

3. Bonanza Creek Experimental Forest, Alaska
   Taiga

Research topics: Successional processes associated with wildfire and floodplains; facilitative and competitive interactions among plant species throughout succession; plant-mediated changes in resource and energy availability for decomposers; herbivorous control of plant species composition

4. Cedar Creek Natural History Area, Minnesota
   Eastern deciduous forest and tallgrass prairie
   Research topics: Successional dynamics; primary productivity and disturbance patterns; nutrients budgets and cycles; climatic variation and the wetland/upland boundary; plant-herbivore dynamics

5. Central Plains Experimental Range, Colorado
   Shortgrass steppe
   Research topics: Soil water; above- and belowground net primary production; plant population and community dynamics; effects of livestock grazing; soil organic matter accumulation and losses, soil nutrient dynamics; and ecosystem recovery from cultivation.

6. Coweta Hydrologic Laboratory, North Carolina
   Eastern deciduous forest
   Research topics: Long-term dynamics of forest ecosystems including forest disturbance and stress along an environmental gradient; stream ecosystems along an environmental gradient; and the riparian zone as a regulator of terrestrial-aquatic linkages

7. Harvard Forest, Massachusetts
   Eastern deciduous forest
   Research topics: Long-term climate change, disturbance history and vegetation dynamics; comparison of community, population and plant architectural responses to human and natural disturbance; forest-atmosphere trace gas fluxes; organic matter accumulation, decomposition and mineralization; element cycling, fine root dynamics and forest microbiology

8. Hubbard Brook
   Experimental Forest, New Hampshire
   Eastern deciduous forest
   Research topics: Vegetation structure and production; dynamics of detritus in terrestrial and aquatic ecosystems; atmosphere-terrestrial-aquatic ecosystem linkages; heterotroph population dynamics effects of human activities on ecosystems.

9. Jornada Experimental Range, New Mexico
   Hot desert
   Research topics: Desertification; factors affecting primary production; nitrogen cycling; animal-induced soil disturbances; direct and indirect consumer effects; organic matter transport and processing; vertebrate and invertebrate population dynamics.

10. Kellogg Biological Station, Michigan
    Row-crop agriculture
    Research topics: Ecological interactions underlying the productivity and environmental impact of production-level cropping systems; patterns, causes, and consequences of microbial, plant, and insect diversity in agricultural landscapes; gene transfer, community dynamics, biogeochemical fluxes:

11. Konza Prairie Research Natural Area, Kansas
    Tallgrass prairie
    Research topics: Effects of fire, grazing and climatic variability on ecological patterns and processes in tallgrass prairie ecosystems; use of remotely sensed data and geographic information systems to evaluate grassland structure and dynamics.

12. Luquillo Experimental Forest, Puerto Rico
    Tropical rainforest
    Research topics: Patterns of and ecosystem response to different patterns of disturbance; land-stream interaction; management effects on ecosystem properties; integration of ecosystem models and geographical information systems

13. McMurdo Dry Valleys, Antarctica
    Polar desert oases
    Research topics: Microbial ecosystem dynamics in arid soils, ephemeral streams, and closed basin lakes; resource and environmental controls on terrestrial, stream and lake ecosystems; material transport between aquatic and terrestrial ecosystems; ecosystems response to greater hydrologic flux driven by warming climate

14. Niwot Ridge-Green Lakes Valley, Colorado
    Research topics: Patterns and controls of nutrient cycling; trace gas dynamics, plant primary productivity and species composition; geomorphology, and paleoecology

15. North Temperate Lakes, Wisconsin
    Northern temperate lakes; eastern deciduous forests
    Research topics: Physical, chemical and biological limnology; hydrology and geochemistry; climate forcing; producer and consumer ecology; ecology of invasions, ecosystem variability; lakescape and landscape ecology

16. Palmer Station, Antarctica
    Research topics: Oceanic-ice circulation and model; sea-ice dynamics; biological/physical interactions; effect of sea ice on primary production, consumer populations and apex predators; bio-optical models of primary production, spatial distribution and recruitment in consumer population dynamics and reproductive ecology

17. Sevilleta National Wildlife Refuge, New Mexico
    Intersection of subalpine mixed-conifer forest/meadow, riparian forest, dry mountainland, grassland, cold desert, hot desert
    Research topics: Landscape/organism population dynamics, watershed ecology; climate change; biospheric/atmospheric interactions; paleobotany/archeaology; microbial role in gas flux; landscape heterogeneity; spatial/temporal variability

18. Virginia Coast Reserve, Virginia
    Coastal barrier islands
    Research topics: Holocene barrier island geology; salt marsh , geology, and hydrology; ecology/evolution of insular vertebrates; primary/secondary succession; life-form modeling of succession

The LTER network office

The LTER Network Office, located at the University of Washington in Seattle, is supported by the National Science Foundation (NSF) to facilitate the achievement of overall Network objectives and initiatives identified by NSF and executive and coordinating committees representing the LTER sites. In addition to coordinating regular meetings of site representatives, the Network Office organizes and facilitates workshops, national and international meetings, and All Scientists meetings.

This information is extracted from the brochure "LTER - National Research Sites with a Common Commitment". Additional information on research programs and collaborative research opportunities may be obtained from:

LTER Network Office
University of Washington
College of Forest Resources AR-10
Seattle, Washington 98195
USA
Tel: +1 206-543-4853
e-mail: Office@LTERnet.edu

NSF, Division of Environmental Biology
Long-Term Projects in Environmental Biology
Tel: +1 202-357-9596
e-mail: jCallaha@nsf.gov

3.7 National Oceanic and Atmospheric Administration: Mussel Watch Project, USA

National status and trends program

Since 1984, the National Oceanic and Atmospheric Administration (NOAA) has monitored, through its National Status and Trends (NS&T) Program, the concentrations of organic compounds and trace metals in bottom-feeding fish, shellfish, and sediments at almost 300 coastal and estuarine locations throughout the United States. The programme was established as a result of increasing public and scientific concern about the quality of the marine environment and the absence of any long-term national monitoring program in the United States. The objective of the program, which is administered by the Coastal Monitoring and Bioeffects Assessment Division of the Office of Ocean Resources Conservation and Assessment, is to determine the status and long-term trends of contamination in these important areas. Samples collected annually through the program are analyzed to determine levels of synthetic chlorinated compounds (e.g., DDTs), polychlorinated biphenyls (PCBs), polynuclear aromatic hydrocarbons (PAHs), and trace metals (e.g., mercury and lead). NOAA's NS&T Program is the first to use a uniform set of techniques to measure coastal and estuarine environmental quality over relatively large space and time scales. A "specimen bank" of samples taken each year at about 10 percent of the sites is maintained at the National Institute of Standards and Technology for future, retrospective analyses. A related program of directed research is examining the relationships between contaminant exposures and indicators of biological responses in fish and shellfish (i.e. bioeffects) in areas that are shown by the NS&T monitoring results to have high levels of toxic chemicals.

Recent trends in coastal environmental quality: Results from the first five years of the NOAA Mussel Watch Project

Introduction
The Mussel Watch Project is a major component of the NS&T Program. Since 1986, Mussel Watch has been making the same chemical measurements on surface sediments and whole softparts of mussels and oysters collected from about 200 coastal and estuarine sites. Recent results from the Mussel Watch Project describing the spatial distribution of coastal contamination and where temporal trends exist, show contamination to be decreasing in many instances. This finding implies that some benefits have resulted from the management of chemical use and discharge. However, data will be necessary for more years to distinguish the effects of human activity from those of natural influences on some of these chemical concentrations.

Sampling sites and species
The need for large-scale and long-term monitoring was emphasized by a U.S National Research Council report (1990) indicating that more than $130 million is being spent every year on U.S. marine environmental monitoring, but that most of it is devoted to compliance monitoring, i.e., testing wastewaters and other materials prior to discharge, or making measurements near discharge points as prescribed by regulation. Since compliance monitoring, by design, covers very small spatial scales, national programs such as NOAA's NS&T Program are the only ones focusing on wider public concerns. It is on this wider scale that national benefits should be derived from spending billions of dollars to control direct and indirect chemical discharges to coastal and marine waters.

The Mussel Watch Project was designed to describe chemical distributions over national and regional scales. Therefore, it is important for sampling sites to be representative of large areas rather than the small-scale patches of contamination commonly referred to as "hot spots". To this end, no sites were knowingly selected near waste discharge points. Furthermore, since the Mussel Watch Project is based on analyzing indigenous mussels and oysters, a site must support a sufficient population of these mollusks to provide annual samples.

NS & T sampling sites are not uniformly distributed along the coast. Within estuaries and embayments, they average about 20 kilometers (KM) apart, while along open coastlines the average separation is 70 km. Almost half of the sites were selected in waters near urban areas, within 20 km of population centers in excess of 100,000 people. This choice was based on the assumptions that chemical contamination is higher, more likely to cause biological effects, and more spatially variable in these waters than in rural areas.

In 1986 and 1987, 145 Mussel Watch sites were sampled. In 1988, a few sites were added on the East Coast to fill in large spatial gaps between sites, and one was added in Hawaii to provide a third sampling site for an oyster species that is not sampled elsewhere. Also in 1988, 20 new sites were selected in the Gulf of Mexico for the specific purpose of gathering samples closer to urban centers.

Results from the initial sampling showed that the highest chemical concentrations were near urban areas on the East and West Coasts, and that few sites in the Gulf of Mexico could be considered contaminated. Since urban centers along the Gulf are further inland than those near other coasts, an attempt was made to sample as close to them as possible. The major limitation on sampling further inland is that oysters are not found at salinities below about 10 parts per thousand. By 1990, 234 sites had been sampled, with further additions made to test the representativeness of earlier sites.

Chemicals measured
The NS&T program monitors concentrations of trace metals and organic compounds. With the exception of chlorinated organic compounds such as DDT and PCB, which exist entirely as the result of human activities, a certain natural concentration of chemicals exists in mollusks even in the absence of human activity. Chemical concentration exceeding natural levels should be considered "contamination", but the exact line demarcating natural concentrations from contamination is not easily drawn. It depends on the species of mollusk itself as well as on many local and regional conditions.

This information is extracted from the brochure "Mussel Watch: Recent trends in coastal Environmental Quality". Additional information is available from: Thomas P. O'Connor, Chief, Coastal Monitoring Branch, NOAA N/ORCA21, 1305 East West Highway, Silver Spring, MD 20910. USA.

3.8 Terrestrial Ecosystem Research Network (TERN), Germany

Global changes, which include climatic changes as well as changes in land use, can have ecological effects. The quantitative recording, prognosis and ecological evaluation Of these effects, in combination with the development of strategies to avoid and restore damage to ecosystems, are a main field of study of the five German Ecosystem Research Centres, located in Bayreuth, Göttingen, Halle/Leipzig, Kiel (cf. Chapter 2.2) and Munich. These centres, funded by the Federal Ministry of Education, Science, Research and Technology (BMBF), are autonomous scientific units, working on different aspects of human and environmental impacts on terrestrial ecosystems. To coordinate the work on Global Change in Terrestrial Ecosystems (GCTE) of the centres and to contribute to GCTE activities in a highly coordinated approach "TERN" (Terrestrial Ecosystem Research Network of Germany) was founded in 1992 by the centres. Chairmen of TERN are the national GCTE coordinators. Effects of increasing CO_2 concentrations on plants and terrestrial ecosystems. The main goals are to determine and predict the effects of elevated CO_2 on temperate forest ecosystems and on agricultural ecosystems and crop rotations. This is a contribution to GCTE Tasks 1.1.1, 3.1.1 and 3.1.3. The project is divided into a number of topics:

– Determination of the responses of single plants or leaves. This work will include studies of (i) the gas exchange of plants in CO_2 enriched environments and in the presence of other gases like O_3, SO_2 and NOx, (ii) the growth of different plants and genotypes under conditions of CO_2 enrichment and at different locations and (iii) the physiology and biochemistry of elevated CO_2 in a network of research laboratories throughout Germany.

– Studies of gas exchange in today's vegetation. Included in this task are (i) studies of CO_2 exchange in ecosystems at present CO_2 levels in relation to soil conditions and along climatic gradients using eddy correlation techniques and (ii) development of ecosystem carbon models.

– Responses of terrestrial ecosystems to elevated CO_2. These field studies of whole ecosystems under a CO_2 enriched atmosphere will be undertaken using FACE technology on two systems: a mixed forest plantation and an agricultural rotation including wheat, corn and potatoes.

A number of sites have been considered for the third task, the FACE experiments. It has been decided that the forest experiment will be carried out in north-east Bavaria in a young, even aged, mixed Fagus/Picea forest with a closed canopy about 10 m in height. The mixed forest would also allow predictions on the competitive interactions under elevated CO_2 levels of broad leaf and conifer species.

The agricultural FACE experiment will be probably carried out close to Munich. The experiment will be part of a large scale investigation of nutrient cycling in agricultural crops under different intensities of agricultural management. The crops wheat, corn and potatoes will be grown in rotation.

The FACE experiments will employ two rings, each with CO_2 concentrations controlled at 650 ppm. Detailed studies of nutrient cycles will be carried out using ^{13}C, ^{15}N and ^{34}S labelling techniques. Trace gas fluxes will be measured at elevated and ambient CO_2 concentrations.

This project also contributes to the CLIMEX experiment of the EC, in which an entire watershed in Norway is enclosed by a greenhouse for studies on elevated CO_2. The system has previously been used for acid rain research.

Regulation of the energy and water exchange of vegetation surfaces

The aim is to develop the capability of predicting the effects of vegetation changes on water and energy fluxes between land surfaces and the atmosphere. In particular the changes of bulk surface conductance with season, succession and long-term CO_2 increase will be studied.

This is a contribution to GCTE Task 1.3.1.

The project has been designed to quantify bulk surface conductance and to determine how global change will affect water and energy fluxes at the patch and landscape scale. The objectives of the project are to (i) determine canopy conductances for different plant communities, (ii) investigate the regulation of water and energy exchange by climate, soil and soil conditions, (iii) account for patchiness, (iv) develop or improve canopy conductance and SVAT models based on plant physiology, climate and soils, (v) validate SVAT, soil water and catchment models using long-term observation plots of agricultural and forest ecosystems along a climatic gradient.

The project is divided into a number of topics:

– Measurement of canopy transpiration. Xylem flow techniques will be used to measure seasonal transpiration in coniferous and deciduous canopies. The results will be compared with measurements using eddy correlation techniques and with soil water models. The work will be used to investigate the effects of patchiness (e.g. forest edge effects).

– Transpiration of agricultural crops. A variety of methods – soil water budget, isotope and micro meteorological – will be used to estimate the canopy transpiration of agricultural crops. The work will include both daily and seasonal water flux measurements of crops commonly used in rotations. Effects of slope aspect and inclination, as well as advective energy inputs will be taken into account. The field measurements will be compared with estimates obtained using spectral vegetation indices and other remote-sensing techniques.

– Development of SVAT models. A large number of SVAT models are already in operation. This work will be based on the present state of SVAT modelling in three specific areas: (i) Development of models on a range of scales (ii) development of regional SVAT models driven by remotely-sensed data (iii) validation of SVAT models using long-term observation plots.

– Use of existing models. Soil water models, SVAT and catchment models are used to evaluate the consequences of global change on water and energy budget. Special attention will be paid to the linkage of SVAT and soil water models to conventional hydrologic catchment models.

– Evaluation of existing data sets. Existing hydrologic data sets of extreme years will be used to test and calibrate models.

The programme will be closely linked to BAHC tasks.

Effects of global change on the biogeochemical cycles of C, N, S and P

The main goals are to (i) quantify the C, N, S, and cation cycles in coniferous and deciduous forest ecosystems as well as in agricultural systems on different soil types and different climatic conditions, (ii) measure the effects of elevated soil temperatures on the biogeochemical cycles in field experiments, (iii) investigate the transformation of N in forest trees and soils along an environmental gradient ranging from northern Sweden (Umea) to southern France (Montpellier), (iv) study the decomposition of soil organic matter and plant residues, (v) investigate the C, N and S transformations using stable isotopes in different soils under different management and climatic conditions and (vi) model C, N and S turnover with modified SVAT models.

This project contributes to GCTE Activity 1.2 and Activity 3.1.

This project is based on a long history of biogeochemical research in Germany. One of the first studies of nutrient budgets started in 1966 in the Solling forest. This site has been studied continuously since then, yielding important information on acid deposition and soil acidification processes. The Göttingen Ecosystem Centre maintains several forest sites in northern Germany for studies of biogeochemical cycles. Detailed research on nitrogen transformations in soils is being carried out at Bayreuth, while Kiel is undertaking landscape-level studies of the interactions of terrestrial ecosystems with surface waters (Cf. Chapter 2.2).

This project is closely linked to a number of existing EU projects: (i) CORE: Exchange of soil columns and exposure to different climates - Munich; (ii) DECO: Decomposition studies of standard plant materials – Göttingen and Munich; (iii) EXMAN: Manipulation of nutrient cycles by exchanging acid rain with clean rain and by imposing drought (rain out shelter) – Göttingen; (iv) NITREX: Study of nitrogen-saturated forest ecosystems – Göttingen; (v) ENCORE: European network of watershed studies – Göttingen (management of data bank); (vi) NiPhys: Study of nitrogen transformation (Bayreuth).

This project is composed of a number of tasks, including both experiments and modelling:

- Input/output budgets of forest and agricultural ecosystems. This experimental task includes measurements of nutrient budgets in watersheds, in deciduous and coniferous forests, and in agricultural systems and watersheds under conditions of crop rotation.

- Input/output budgets in soils with 3°C higher temperatures in the topsoils. The change of fluxes and processes on the heated experimental plots will be investigated involving all TERN groups.

- Transformation of nitrogen in trees and soils. This task is part of the Nitrogen Physiology (NiPhys) project which is coordinated by Bayreuth in the framework of EC research. Sites range from northern Sweden to southern France, but only the Bayreuth contributions to NiPhys are listed here: investigation of nitrate reductase activity; measurement of ^{15}N values along a gradient of pollution; denitrification in relation to climate and soil conditions; adsorption and immobilization of deposited nitrogen.

- Changes in C, N, and S turnover in transplanted cores of agricultural soils of different origins along a climatic transect through Europe. Work within this task includes: decomposition patterns of organic soil matter; decomposition of ^{13}C, ^{15}N, and ^{34}S labelled plant residues; changes in microbial and faunal activities; gaseous and leaching losses of C, N and S compounds.

- Management effects on C, N and S turnover in agricultural soils (tracer studies). The components of this task are: effects of reduced tilling on C, N and S dynamics in soils; organic and inorganic fertilizer effects on C, N and S dynamics; management effects on gaseous losses and leaching; effects of rotation and intercroppinq on C, N and S turnover.

- Modelling. This task will build on the experimental results of the other four tasks to develop models of nutrient cycling which include canopy processes of water and energy exchange, and to develop regional models of landscape evaporation.

This project is well placed to link with GCTE's biogeochemical transect in Scandinavia (Task 1.2.2 – Changes in Biogeochemistry in High Latitude Systems) to provide a transition to the temperate forest biome. An extension further south to France is also possible.

Emission of trace gases from terrestrial ecosystems

The main goal is to determine the release and uptake of trace gases, above all CO_2, N_2O and CH_4, from soils of different terrestrial ecosystems, i.e. forest ecosystems, agroecosystems and wetlands. The investigations are intended to demonstrate the spatial and temporal variability of the gas fluxes. The importance of land use strategies and environmental factors (soil temperature and water content) to the gas emission will be examined in additional laboratory experiments.

This project contributes to GCTE Task 3.3.3.

On account of the extraordinary position of C and N in the element budget of terrestrial ecosystems the investigations will be focused on gaseous C and N compounds, starting with CO_2 and N_2O. The main emphasis will be placed on the monitoring of trace gases in dominating central European ecosystems, i.e. forest and agroecosystems. Some neighbouring systems, e.g. wetlands, will be included in the studies.

Further, we hope to improve our knowledge of the interrelations of trace gas exchange between the biosphere and the atmosphere and the key factors (land use, soil biology, chemistry and physics) will be improved. Laboratory studies will be conducted under controlled conditions.

The influence of higher soil temperatures (a consequence of the rising temperature forecast in central Europe) on trace-gas emissions will be investigated by linking up with the field experiment of the biogeochemical TERN group (iii).

The following main hypotheses are the backbone of the working schedule for all research groups participating:
- Emissions of trace gases from soils are mainly influenced by land-use strategies. Therefore forests and agroecosystems show differing, typical patterns of gas release.

- Trace-gas emissions are significantly altered by different atmospheric deposition inputs of S and N. Also N fertilizers, especially mineral ones, change the gas fluxes by their amount and method of application. N_2O increases significantly due to these effects.

- Trace-gas emissions can be estimated by ascertaining the water, air and temperature regimes of soils, the chemical composition of the soil solution and the biological activity of the soil.

- The fluxes of N_2O and CO_2 are significantly increased by an increase in soil temperature of $2°-3°C$.

- The quantification of trace-gas fluxes enables us to demonstrate their importance to the ecosystem balance of C and N.

The programme will also contribute to IGAC Activity 7.2.

Scaling-up of ecosystem processes and development of a TERN information system

The main goal is to test and develop theoretical approaches for the scaling-up of ecosystem processes. This is combined with the elaboration of concepts and methods to adapt models and data aggregation procedures to different regional scales. This project integrates TERN activities and develops procedures for regionalization.

The project comprises the following components:

1. Testing and elaboration of theoretical limitations of different scaling procedures. Special emphasis will be placed on the analysis of the hierarchy theory in order to describe scaling-up problems. With respect to the levels of investigation in the whole TERN approach different scales can be tested.

2. Based on the other four TERN activities the methods and data available will be used in models which differ in complexity in time and space. By means of parallel investigations on the different scales it will be possible to develop methods of data aggregation and procedures to adapt models with different resolutions in time and space. Of special interest is the identification of interactions between the different levels under investigation.

The following hypotheses will be tested:
- Increasing CO_2 concentrations (and temperatures) will cause different changes of the ratio between fixing and mineralization of C on different scales.

- The expected increase of CO_2 and temperature will change the interactions between the organizational levels in an agricultural landscape.

- In heavily-loaded forest ecosystems the controlling capabilities of higher hierarchical levels is weakened. Changing climatic conditions in such ecosystems will cause stronger changes on the different scales than in slightly-loaded forest ecosystems.

- Changes in land-use patterns caused by global change will alter water and biogeochemical cycles on the watershed and regional levels.

- Socio-economic constraints and agricultural management practices are able to overrule changes caused by altered climatic conditions.

This information is extracted from the brochure 'IGBP Research in the Federal Republic of Germany' (1993), with the kind permission of the IGBP Secretariat, Berlin, Germany. Additional information about TERN can be obtained from GSF-Forschungszentrum für Umwelt und Gesundheit, Institut für Bodenökologie, Neuherberg, Postfach 1129, D-85758 Oberschleißheim, Germany, Tel: +49 89 3187 4064, Fax: +49 89 3187 3376.

3.9 Chinese Ecological Research Network (CERN), China

Background

In 1987, the Chinese Academy of Sciences (CAS) established a Chinese Ecological Research Network (CERN) by linking its established ecological and environmental observation and experiment stations located at different natural zones across the country. CERN is intended to help understanding of the mutual relationship between ecosystems and their impact on the environment and enable long-term observation of their status and characteristics. It will contribute to global change programs such as IGBP and the global observation network of ecosystems and environments.

A Network Leading Group was organized by CAS, to investigate to accomplish CERN. A final experiment was planned and approved in 1989. For this 21 stations were chosen to conduct network research with unified methods. The main contents of the network research include:
- Demonstration research on structures, functions of main types of ecosystems in China and the development of stable ecosystems with high productivity.
- Experiment study research on critical processes and the long-term impact of human activities on ecosystems.
- Long-term observation and monitoring on ecosystem evolution and environmental changes.

Based on these main research items, eight observation-experiment networks were formed as follows:
- Demonstration of optimized model of ecosystems (in 17 stations)
- Crops water demand and consumption (in 7 stations)
- Decomposition, accumulation and equilibrium of organic matters in soil (in 5 stations)
- Release of soil potential nutrients and their dynamics (in 3 stations)
- Methane emission from paddy fields (in 3 stations)
- Land loading capacity and the radiation spectrum (in 8 stations)
- Long-term impact of different agricultural operation systems on environment and ecosystem productivity (in 5 stations).
- Long-term impact of various ploughing methods on soil environment and land productivity (in 2 stations)

Technical regulations for field observations and experiments were set up and a strategic planning of CERN information systems was completed.

In 1991, the Chinese government authorized CERN as a national construction project of its 'Eighth Five-Year Plan' and authorized CERN to apply for a World Bank loan.

Key objectives and features

The characteristics of CERN will be as follows:
1. CERN will be an interdisciplinary and comprehensive research network. CERN is not only to conduct long-term monitoring of ecological and environmental elements, but more importantly, is also to carry out network wide studies on structures and functions of ecosystems, and to reveal the mechanism of ecological processes through long-term site experiments. These experiments take into account the fact that ecosystems comprise both biotic

and environmental subsystems. Long-term observations and experiments in the 6 subsystems will be analysed to study trends of changes in the structures and functions of the systems. In light of the social and economic conditions of the areas where the stations are located, CERN will propose coordination/control and management of the ecosystems, and to set up optimized demonstration models by close cooperation with the local governments and people.

2. CERN will focus on studies of the trends of changes in structures and functions of ecosystems as affected by intense human activities. This fits the national conditions of China. China is a developing country with a rapid growth of population and has resources per capita far below the world average. Resources are unevenly distributed, and have been irrationally developed and utilized for a long time. This has caused many ecological and environmental problems, such as land desertification, soil erosion, and aggravated pollution. Therefore, it is an important task for CERN to constantly explore approaches for setting up new artificial complex ecosystems (natural and social environments and human beings) in order to realize the unification of social, economic and ecological benefits. Natural ecosystems are also an aspect of CERN research since long-term stationary studies on structures and functions of natural ecosystems can contribute to understanding the mechanism of processes and the law of evolution that are advantageous to direct the building of optimized artificial ecosystems.

3. CERN will use its network of stations to predict macroecological environmental problems caused by human activities. This part of Earth-system research has a global focus as well as scientific and social significance. The main steps of this Earth-system research are to observe, to analyze the data, to establish conceptual and numerical models and to evaluate the models and make predictions.

A measurement index system will be one of the critical factors of CERN construction. It will decide not only the content of the long-term stationary observations and experiments and the data generated but also technical configurations of the network. CERN has defined an index system for observation variables. The system will comprehensively reflect the structures of ecosystems and the physical, chemical and biological processes in the transmission of energy, nutrients and water. An index matrix was formulated from the point of view of integrity of ecological data. The columns of the matrix represent structure and essential functional factors, while the rows refer to different physical systems. There are interfaces between any two of the systems. The variables at these interfaces have been defined by interactions with IGBP. The CERN index system will be implemented in steps.

Structure

The structure of CERN forms a network with three tiers: ecological research stations, subcenters, and a Synthetic Center at its top. Four support systems, which will shape CERN as a matrix-like system, are research system, information system, technical system, and management system.

CERN information systems

In order to predict the trends of changes in ecosystems and the environment and to develop the strategies and technologies for regulation and control, it is necessary to provide the long-term data related to environmental changes and ecosystem successions and data management. Therefore, the establishment of data and information systems has become an important task of CERN.

The information system CERNIS will be a distributed system in three tiers managing the following six data sets as its main functions:

- Data 1: reflecting the results from the long-term site observations on the basic ecological measurements, including meteorological factors, hydrological factors, soil factors, biomass production, and the qualities of air, water, soil, and biotic environment, which are required to study the structures and functions of ecosystems in an ecological station.

- Data 2: reflecting the information required to study the processes of ecosystems. Currently, the long-term stationary experiments are carried out on five processes; i.e. water cycle and equilibrium in soil-biotic-atmospheric system; nutrient substance cycle and equilibrium; energy flow, generation, transport and transformation of trace gases; and decomposition, accumulation and transport of organic pollutants and heavy metals.

- Data 3: reflecting the social and economic indicators of the areas where the ecological stations are located and basic conditions of demonstration areas, as well as research project data catalogue and managerial data at ecological stations.

- Data 4: reflecting the disciplinary information required to study the ecosystem and environment of a medium and large scale in the network, collected and compiled be the four subcenters.

- Data 5: the information on ecosystems, resources and environment of large scale (nationwide, macroscopic) features, collected and maintained by the central information system.

- Data 6: the information from the outside CERN through data exchange.

In order to reinforce data management and data quality control CERN will set up its Data Management Committee and assign Data Administrators in stations, subcenters and the synthetic center. Current efforts include designing data management systems, making a data management training plan, formulating relevant regulations and procedures in data quality assurance and quality control (QA/QC), archiving existing data and information at stations, and planning experimental instrument and facilities.

Conclusion

The development of CERN is strongly supported by the Chinese government. It has been listed in "Eighth Five-Year Plan" as an important development project and is applying for a World Bank loan.

When finally established, CERN will be an ecological research network under the Chinese Academy of Science consisting of 29 stations, 4 subcenters and a synthesis center. This will improved the ecological research in China from a traditional, empirical, and qualitative discipline to a modern ecological science well equipped with contemporary theories, methodologies and technologies. Through networking stations it will be possible to scientifically predict the trends of changes of our living environments and the impact caused by human activities, to provide scientific evidence to national economic development and to make greater contributions in dealing with global environmental.

This information is extracted from a paper prepared for the UNEP-HEM workshop. Some of this information is included in 'Climate Biosphere Interaction: Biogenic and Environmental Effects of Climate Change', Chapter 4, edited by Richard G. Zepp and published by John Wiley and Sons. ISBN 0-471-58943-3. Reprinted by permission of John Wiley & Sons, Inc. Additional information may be obtained from: Zhao Jianping, Bureau of Resources and Environmental Sciences, Chinese Academy of Sciences.

3.10 Environmental Change Network (ECN), UK

Introduction

A long-term integrated monitoring network is being developed in the UK to identify and quantify environmental changes associated with man's activities, distinguishing man-made change from natural variations and trends, and giving warning of undesirable effects. The UK Environmental Change Network (ECN) is operated by a consortium of sponsoring organisations with an interest in land-use and the environment, and is managed by the Natural Environment Research Council. The operation of ECN depends on the voluntary collaboration of these agencies in providing sites and the necessary funding for staff to carry out the monitoring programme which is scheduled to carry on for at least 30 years.

The sites

The UK has many sites with a long history of environmental data collection and repeated surveys. In selecting suitable sites for ECN it was obviously sensible to capitalise, as far as possible, on some of these sites with their known management history, existing data and environmental understanding. A list of twenty-four such sites was compiled, representing the broad range of climate, soil, habitat and land management in the UK It has been possible to recruit ten terrestrial sites into the network so far; it is expected that two others will be committed this year and that others will join later. The sites range from small intensively-managed lowland agricultural establishments to large, semi-natural upland areas.

The measurements

The use of standardised methods of data collection is an important principle of ECN and is to be achieved by using agreed standard protocols for all the measurements to be undertaken. The programme is centred around a series of 'Core Measurements' which are made at all the sites with the exception of one or two which are impossible at some sites. The measurements relate to variables which are expected to b important in driving environmental change, and to ecosystem response variables which have been identified as being sensitive or responsive to such change. Some of the measurements are concentrated on a designated area of one hectare, the Target Sample Site, whilst others are more appropriately carried out on other, and often more extensive, areas or along designated transects.

In addition to recording land-use changes at each site a number of physical and chemical features of the environment are recorded. Meteorological data are collected using both automatic weather stations (AWS) and standard manual recording. The AWS records 12 variables at 5-second intervals and stores the data as hourly summaries which are downloaded fortnightly. Rainwater is collected for regular chemical analysis and passive diffusion tubes are used to collect some pollutants such as NO_2; sites with perennial streams record discharge at a permanently installed weir or flume using a digital logger, whilst weekly dip samples are analysed for major cations and anions. The chemical composition of soil solution is measured fortnightly using suction lysimeters placed at two different depths in the soil. Soils at each site are characterised by survey, and different physical, chemical and mineralogical properties are assessed on two cycles of five and twenty years.

Whilst as many as possible of the physical measurements in the programme are automated the biological part of the programme relies on more traditional methods of data collection. Vegetation monitoring requires as its basis an accurate and comprehensive description of each site's vegetation at the outset. This is achieved by mapping and collection of data from up to 500 sample plots, some of which are temporary; the observed variation is then characterised by reference to an existing descriptive scheme, the National Vegetation Classification, which allows the comparison of change across the network. Monitoring follows at two levels: coarse-grain at intervals of 9 years on at least 50 permanently marked plots chosen at random, and fine-grain monitoring every 3 years at locations in each of the vegetation types established as a result of the earlier vegetation characterisation exercise. Provision is also made for monitoring species composition of linear features such as hedgerows and boundaries between vegetation types.

Recording of animals is directed towards groups believed to be good indicators of environmental change and for which there are already good ecological data which can provide a sound background to interpretation. Animals with wide distributions will be used for inter-site comparisons of possible changes with time; those for which there are already national monitoring schemes, into which results from ECN sites can feed, have also been included.

The Common Bird Census (CBC) is an established national system for monitoring birds which are fairly numerous and therefore provide samples large enough to build a population index. Application of this methodology will provide annual counts of territorial birds matched against the main habitat features at each site. At some of the upland sites the CBC is inappropriate and in these circumstances the annual distribution and abundance of moorland breeding birds will be assessed within selected areas. Surveys of changes in rabbit populations should provide information on how changes in weather, disease and land-use affect rabbit numbers. Bat surveys will use a tuneable bat detector operating at 45 kHz to record year-to-year changes in populations of this climate-sensitive mammal. Frogs are known to be sensitive to some pollutants and there is a marked geographical trend in their spawning dates whose pattern may be affected by climatic changes.

A number of invertebrate groups have been included in the programme. Moths (macrolepidoptera) and butterflies already have national monitoring programmes and ECN sites will follow the existing methods of these surveys and contribute data to their programmes. This will allow ECN sites to be placed in a regional and national context. Two species of spittlebug (Philaenus spumarius and Neophilaenus lineatus) are widespread and common throughout the British Isles and the programme will both estimate nymph densities and also determine the proportions of adult colour morphs because it is likely that these proportions of these are environmentally sensitive and will be a good indicator of environmental change. Other groups, such as craneflies and ground beetles, will also be monitored

Development of the ECN database

During the initial phase of the programme there have been three principle areas of work in relation to the development of the ECN Database:

1. Database design
2. Design of standard data handling and transfer formats
3. Software development

The successful operation of the database depends largely on its efficient design. It needs to handle and retrieve efficiently large quantities of data. It must be capable of providing for a wide variety of analyses relating to both space and time dimensions, their respective data characteristics and analytical operators. The structure of the data tables within the database has now been designed and implemented; this includes a meta-information system which will carry essential information about data origins, their spatial and temporal dimensions, and their quality.

Standard data recording forms and transfer formats have been produced for each of the Core Measurements and for the associated quality information, to facilitate data handling at the ECN Central coordination Unit (CCU). The design takes into account ease of field use, data entry and requirements for final structures within the database. Recording forms and transfer formats are complete for most Core Measurements and are now in use at ECN sites. Data are sent from ECN sites to the CCU on disk or via electronic mail, using agreed procedures. An overall Data Handling Protocol describes the general principles to be used for the transfer of data.

The ECN Database uses the relational database management system (RDMS) "Oracle", which is linked with the geographical information system (GIS) "Arc/Info", as its software platform. Software for data input, validation and retrieval is being developed, using Oracle and Arc/Info macro-languages where possible, for each ECN Core Measurement, and for access to the database and to the meta-information system as a whole. Sites are responsible for checking data before sending it to the CCU but validation and corrections are also made centrally at the CCU.

Sites formally began collecting data in January 1993, although many of the biological measurements were necessarily scheduled for spring/summer 1993. Data are now being transferred for those measurements which are to be taken regularly throughout the year and therefore started on or before January 1993. Standard time-schedules for data transfer from sites to the CCU have been specified for each Core Measurement. Although the expected 'teething' problems with instrumentation and organisation at sites in the first year have inevitably caused some delays, these schedules are confidently expected to become routine within a short time.

A set of site maps showing vegetation cover, soils and detailed locations of sampling sites used for Core Measurements are being established within the GIS.

Looking ahead

ECN recording started in 1993, which in some respects is regarded as a trial year during which the protocols will be tested and modified as necessary and other sites will join the network. The next phase in development is the expansion of the network into the aquatic environment and this will gather momentum during 1994. Discussions are taking place between organisations in England, Scotland, Wales and Northern Ireland as a result of which it is expected that up to 16 lakes and 20 rivers throughout the UK will be included; a group of aquatic scientists is drawing up a series of protocols analogous to those used in the terrestrial network. At a later stage the scope of the programme might be extended to include estuarine and marine sites.

ECN is keen to forge links with groups planning to establish similar networks in other countries and will be glad to share its experience with them. A bi-annual newsletter, ECN NEWS, is available to interested readers who contact the author.

Annex II

This information is extracted from a paper submitted to UNEP-HEM for the workshop. Additional information can be obtained from J.M. Sykes, ECN Coordinator and Deputy Head of Station at the Institute of Terrestrial Ecology, Merlewood Research Station, Grange-over-Sands, Cumbria LAI I 6JU, UK.